JOURNALISM IN THE ANTHROPOCENE

Media, Modernity and the World to Come

Dominic Hinde

First published in Great Britain in 2026 by

Bristol University Press
University of Bristol
1–9 Old Park Hill
Bristol
BS2 8BB
UK
t: +44 (0)117 374 6645
e: bup-info@bristol.ac.uk

Details of international sales and distribution partners are available at bristoluniversitypress.co.uk

© Bristol University Press 2026

DOI: 10.51952/9781529256741

British Library Cataloguing in Publication Data
A catalogue record for this book is available from the British Library

ISBN 978-1-5292-5671-0 hardcover
ISBN 978-1-5292-5672-7 paperback
ISBN 978-1-5292-5673-4 ePub
ISBN 978-1-5292-5674-1 ePdf

The right of Dominic Hinde to be identified as the author of this work has been asserted by him in accordance with the Copyright, Designs and Patents Act 1988.

All rights reserved: no part of this publication may be reproduced, stored in a retrieval system, or transmitted in any form or by any means, electronic, mechanical, photocopying, recording, or otherwise without the prior permission of Bristol University Press.

Every reasonable effort has been made to obtain permission to reproduce copyrighted material. If, however, anyone knows of an oversight, please contact the publisher.

The statements and opinions contained within this publication are solely those of the author and not of the University of Bristol or Bristol University Press. The University of Bristol and Bristol University Press disclaim responsibility for any injury to persons or property resulting from any material published in this publication.

Bristol University Press works to counter discrimination on grounds of gender, race, disability, age and sexuality.

Cover design: Nicky Borowiec
Front cover image: estivillml/AdobeStock

In modern industrial societies the primary and most accessible 'social theory' is provided by the media [...] Journalists, however, do not consider the production of high quality social theory and analysis as one of their primary tasks.
Barbara Adam

Contents

Preface		vi
1	Journalism, Modernity and the Human Earth	1
2	The End of Nature and the End of Journalism	13
3	The Anthropocene in Media	31
4	Media Hot and Hotter	41
5	Plotting the Anthropocene: Digital Modernity and Charismatic Data	51
6	Reporting the Climate City	66
7	New Times for Journalism	79
8	Resilience and the Never-Ending Shock of the Present	89
9	Media Cosmopolitanism in the Anthropocene	99
10	Global Journalists in Liquid Worlds	112
11	Journalism as Eschatology and the World to Come	122
Notes		132
References		133
Index		143

Preface

This book very nearly did not exist at all, and is very different to the book that I originally intended to produce. When I first sat down to work on the manuscript at the Rachel Carson Center in Munich I had a brief and a grant to write a book about climate change, and I expected it would take me a few months to pull together from my notes. What I had not anticipated was that just weeks later I would wake up on the pavement after a bike accident with injuries that would stop me using a computer, talking at length or writing well and coherently for years afterwards as my brain adjusted to the impact of head trauma. Simply getting the words on the page in a coherent order represents something of a victory, but five years later the book has also morphed into something quite different from what I initially envisaged.

In the end one book turned into two, and while the other book went in a more creative direction, *Journalism in the Anthropocene* owes its contents to a short study I was funded to carry out by the Institute for Advanced Study in the Humanities at the University of Edinburgh just after I returned to academia from full-time journalism. Lots of lying in dark rooms, however, and the twin impacts of brain injury and a subsequent COVID-19 infection let me rethink some of the findings from my time in Edinburgh and add new content and research, trying to think through theory in the live laboratory of the COVID pandemic and the shocks to the world system it heralded.

Thankfully, as the pandemic ended I was able to move to a new supportive research environment as part of the Glasgow University Media Group, in a sociology department among like-minded colleagues, and I owe huge thanks in particular to Professor Matt Dawson who took a gamble on me in bringing me to Glasgow. It is very hard for me to read through the manuscript without seeing everything that has happened in between on its pages, and hopefully rather than being overtaken by events it is clear how this book has been formed by the processes going on around it.

There are a few moments in particular that have stuck with me from the six years it took to finish what I started, and that reminded me of why I first became a journalist almost two decades ago. Sitting on the shingle beach by the film-maker Derek Jarman's cottage on the Dungeness peninsula in the English Channel, I witnessed a migrant boat full of freezing displaced people being hauled ashore and shipped off to a holding centre by UK Border Force personnel. It was a timely reminder of the powerlessness we can feel as both journalists and audiences when confronted by global processes intruding on our relatively comfortable lives, but also one of the first things I wrote about for a commercial outlet since becoming ill. When the boat appeared out of the fog that day I started taking pictures and making notes, not knowing what I would do with them but sure there was some good to doing so as an uninvited witness. Journalistic responses are often very different to academic ones; responsive, ad hoc and unplanned, even opportunistic. Both have a place in helping us to better grasp where we are and how the world moves around us, as well as how we move it.

Then, just as I was doing the final edits, I happened to be nearby and was asked to report on the flash flooding in the suburbs of Valencia in Spain, wading through mud as tens of thousands of people were displaced by extreme storms. Standing in the abandoned underground car park of a furniture superstore with cars flipped over and water around my ankles, climate change has never felt more real. I interviewed volunteers handing out food parcels and people whose whole lives had been destroyed by just a few hours of rain. This book is about how we convey these moments and give them meaning.

Along the way I have borrowed ideas from my fellow-travellers in environmental anthropology, called on my own experiences over the past twenty years as a foreign and environmental journalist, and sat down with climate scientists, politicians and other media workers to get their opinions on where they fit into the mediation of climate change and environmental transformation. I owe a huge debt to a large number of colleagues in journalism in particular who have had off the record conversations with me. This kind of informal research has a long tail, and I hope that you recognise how your own experiences have been reflected in the final work.

No interrogation of climate is possible without also investigating the evolving structures of the modernity that created climate change, and by extension journalism's place within it. The challenge of public understanding and political action is intimately related to how media grasps this larger unfolding and articulates its future possibilities. At its root, perhaps, this book is about how the media perpetuates ideas of time and effect, but also about the fragile balance of structured grand theory and everyday circumstance that runs through the social sciences and public media alike.

The issues at the heart of this book are not going to go away or lose relevance, and I hope that in its own way it might contribute to making the coming decades a little easier to navigate.

Marseille, France
December 2024

1

Journalism, Modernity and the Human Earth

> At the moment of Sputnik the planet became a global theatre in which there are no spectators but only actors.
>
> Marshall McLuhan

On 26 June 1967 Marshall McLuhan jumped into the sitting rooms of TV viewers across the planet as the face of *Our World*, the earth's first simultaneous international live television broadcast. Most of the viewers who saw McLuhan being interviewed in the Toronto studios of the Canadian Broadcasting Corporation would have had little idea who the talking head of the Information Age was, were it not for the formal introduction on screen. Yet McLuhan's presence gave academic weight and – crucially – some sense of context to the audacious attempt to capture modernity's triumphs and risks in a singular global moment. *Our World* took audiences out of their bodies on a real-time journey around the globe through vignettes of life in participating nations. The Beatles played *All You Need Is Love* live from a London studio, Japanese labourers were filmed undertaking subway construction in Tokyo, while audiences got a live taste of cattle farming, modern art galleries and deep space tracking stations. At the heart of the broadcast was the conceit that the goings on of life and human history were not merely facts repeated and printed on the pages of books and newspapers or regurgitated on the evening news, but events unfolding perpetually just as our own lives tick from second to minute to hour.

A leitmotif in *Our World* was the filming of maternity wards and the lingering question of what type of planet the babies lying in their hospital cradles that June day could expect to inherit. The main studio in London was adorned with contemporary 1960s renderings of da Vinci's Vitruvian Man and a spinning globe prop aimed at communicating the universality of the human experience, night turning to day and back to night under the watchful eye of the orbiting satellites that enabled the whole spectacle.

The script for the show itself was peppered with references to 'mankind' and the global community of peoples, though due to political tensions the Warsaw Pact countries and other communist states were conspicuous by their absence. The final line of the broadcast stated obliquely that 'all of us have to live on our world' as the studio model of the earth slowly rotated on its bearings. The rubric OUR WORLD ran across the screen in multiple languages as the last lines of the script were delivered, and for a few hours of grainy transmission audiences basked in a fledgling global media commons.

Inherent to this celebration of global modernity was the recognition of its dark underbelly: population explosion, nuclear conflict and the spectre of an oncoming wave of technology that carried both promise and danger. Humanity – so the experiment claimed – was at a technological threshold that was transforming the planet. As McLuhan (1974) would go on to write in his discussion of satellite technology and global communication networks, space-borne communications represented a rewriting of the public understanding of the globe and its disparate spaces – *Our World* made it clear not just that there were things going on all around the planet, but that these things were happening in the same interdependent moment and that all of us were actors in this theatre. It was, as the media scholar Lisa Parks has described the event in her work on media and satellite communications, an undertaking that took advantage of 'the unique properties of live satellite television: its capacity to craft a "global now"' (Parks 2005: 22).

According to the parameters set by the broadcast, by participating in modernity we are able to participate in global life through the web of orbiting satellites that began with the Soviet Union's *Sputnik* and continues today with strings of broadband relay stations, enabling online gaming, video calls, and peer to peer commerce, in high orbit or buried on the seabed.

Given the limits of technology in 1967 compared to today, this vision of the world as a participatory theatre seems more than a little hubristic. Nevertheless, the evolution of global broadcasting into online news and social media carries with it the same dream of globality through telecommunications, and these technologies are critical in understanding what has become known as the 'participatory condition' (Barney et al 2016) that governs our social life, promising emancipations while also creating new tools to enforce social structures. McLuhan's vision of a global ecosystem of media and interaction, in which the world and its occupants become enmeshed in an entangled and ultimately enlightened commons, was reflected in the cautious optimism of the global journalism that underpinned *Our World*. Its carefully choreographed vignettes showcased networks of opportunity and threat that attempted to articulate globalisation for a popular audience.

It also framed modernity itself as a difficult but necessary negotiation of the future towards a point of betterment, facing down the inevitable risks and challenges that progress brings as part of the grand narrative of humankind. The human earth of McLuhan's technological utopia is one in which natural global systems are overlaid by an electronic shadow, running from its roots in people's homes and offices into shared rivers and oceans of experience and exchange. Modernity flows and laps on every shore, and *Our World* sought to show how the very modern experience of watching television could connect us to processes and sensations we felt at the peripheries of our existence, but which were also more central to our lives than we might imagine.

Our World was part event television, part educational broadcast, and featured elements of what would become established as the familiar practice of 'going live' in broadcast news. The careful scripting of each section borrowed heavily from the presentational norms of the BBC that had evolved through the 1960s, while also demonstrating the potential for journalism to move from the reporting of recent events to the narration of reality as it unfolded, wherever that may be. No longer was it necessary for the large and small happenings of modernity to be stitched together after the fact.

Journalism, according to the veteran journalism scholar John Hartley (1996), is the primary sense-making practice of this modernity and its flows. It is the frame of reference for both modernity as a structural system and for our reflexive conception of the present and all our future imaginaries. Journalism and the news work as a constantly evolving shared canon and bricolage of the historical moment, a textual system that underpins how we see ourselves and the world we must all live in together. The practice and consumption of journalism is said, in Hartley's words, to 'affect individuals and bring down governments, to expose evils and wreck lives, to be the defence of democracy and the cause of decline, to extend knowledge and spread ignorance, all at once, all over the world' (Hartley 1996: 20).

Making sense of modernity

Now in 2025, almost 60 years on from the first words in a London television studio reaching the homes of *Our World*'s global public, we find ourselves in a similar time of unparalleled knowledge and profound ignorance, menaced by global existential threat and technological advancement which is both seductive and disquieting. How we make sense of these shared problems is contested and uncertain. Journalism, with all its promise and claims to societal importance, is not working in the way its utopian idealists and defenders suggest it might, and neither is modernity at large with all of its transformational energy.

Climate change, entrenched inequality, continued degradation of natural systems and the associated political volatility that follows, have gone hand in hand with a decade-on-decade decline in journalistic capacity that has challenged its institutional centrality to contemporary societies and sown doubts about its ability to serve the public interest at a time of immense need. To look at the news today is a confrontation not only with the growing and distressing currents we see around us, but also with journalism's own ideals and ethics and with its need to justify its existence to audiences and funders alike.

These overlapping crises of journalism and earth systems and their place in modernity are at the heart of this book. Their entanglement with one another provides a lens through which to see the myriad impacts of modernity on the environment and the media system alike, and to understand how both are bound to the Great Acceleration in human development since 1945 and to the idea of the Anthropocene in particular. The Anthropocene has made waves in academia and more modest ripples in the public media, but its calling into existence by earth scientists and humanities scholars alike has failed to have world-changing impacts in the way some hoped. Making sense of the Anthropocene and understanding its power as an idea fundamentally means making sense of the modernity that first created and continues to create it.

Similarly, the question of journalism's future cannot really be fully understood without seeing its place within the infrastructures of modernity, as both a product of and a key factor in global social and technological change. An industry that demands quick answers to what it often self-diagnoses as the most pressing questions, and where media academics are often expected to fulfil the role of consultants and policy advisers focused on questions of efficiency and innovation, can never possibly marshal the disparate and entangled stories that make up journalism as a global object.

At its worst this means academic studies of journalism fall prey to narrow discussions of whatever happens to be in vogue at any particular time, from fake news and AI (artificial intelligence) to questions of representation and visibility divorced from any in-depth understanding of structure and history. Such issues are all of course important points on the compass of contemporary journalism, but in the search for impact and relevance journalism watchers can unintentionally ape the worst of economics' reinvention as a research arm of global finance and political science's annexation by pundits and consultants as a form of data farming and political betting. Watching the media – and indeed working in it – we can too easily become preoccupied with the enchantment of innovation and newness, and an ethic of competition in which there can ultimately be few winners. This is especially true as 'the industry' seeks answers to questions about its own future and self-image through the reassuring weight of an academic letterhead and the funding opportunities that brings, making academics outriders in the push towards a chimeric new journalism of dubious provenance.

This existential doubt in journalism industries is in part a concern about profits and the economic sustainability of a sector that for over a century represented a lucrative investment as well as a direct avenue for political and societal influence. The civic purpose that has become a central ethic of journalism can be seen as part of an emergent system of coproduction rather than a founding principle, yet it is habitually reproduced as journalism's *raison d'être*. Even more, perhaps, journalism is on a teleological quest to find, rebuild and replicate the centrality of journalistic institutions to the public life and politics it helped to co-create. It speaks to a desire to understand and counter the instability and ephemerality of the present and to somehow stop the unravelling of assumptions about the way the world is and the media's place in it.

How we see change matters, and key to this seeing is how we understand the drivers, the beneficiaries and the victims of change, transformation and modernisation. The philosopher Marshall Berman's (1983) description of the experience of modernity as solidity melting into air, reapplying the lens first coined by Marx and Engels to encapsulate the transience and change of Victorian capitalism, seems as true today as it did in the 1840s. In the 1980s when Berman used the idiom to understand contemporary New York and the threads of creation and destruction that were rendered visible in the changing urban landscape, it was also a time of emergent new forms of media capitalism – of news helicopters, infotainment and subscription cable building a new America onscreen and off.

This invocation of Marx and Engels writing from the midst of the European Industrial Revolution is more than a convenient intellectual nod; today's worldwide environmental crises can be traced back over the last two hundred years and to the roots of fossil modernity and extractive resource capitalism, urbanisation, and more latterly the financialisation of our everyday lives, from Manchester to Manhattan and beyond. We are all of us not so much melting into air as being vaporised and reconstituted along with the material world around us.

The original quote from *The Communist Manifesto* that Berman adopted as his central idiom is 'all that is solid melts into air, all that is sacred is profaned, and man is at last compelled to face with sober senses his real conditions of life, and his relations with his kind' (Marx and Engels 2002: 223). This transience, the feeling that just as we think we have adapted to our time we are met head on by new realities and hard truths about the way in which the world has come to be, is not new, but it is critical to finding positive futures amid the insecurity of the present. In the case of the current crisis of the global environment, it is obvious what sober senses and sobering experiences we are confronted with as the real conditions of life begin to impact on our future. Extreme weather and infrastructural collapse lurks in the daily headlines as economic shocks and violent conflicts watch from the wings.

In many ways the popular feeling of the Anthropocene and its uptake as a label for the sensation of a global polycrisis is an attempt to put a name to this experience of our belief in a steady-state and steadfast solidity turning to insecurity under our feet. It is the acceptance that nature is as fluid, fragile and changeable as the human institutions of modernity in constant flux around us. This also entails understanding journalism's role in the experience of this precarious modernity, and how the stories journalism tells, its ability to tell them and its place as a sense-making practice can contribute to escaping or transforming the moment we are in.

Each of the chapters of this book, then, is an attempt to bridge the gap between the work of journalism and the core problematics of the Anthropocene concept and the questions it poses. This comes from the natural sciences, from sociology and from what are broadly termed the environmental humanities alike, showing how the academic labour of the last two decades can be of assistance to journalism, but also how the modes of expression employed by journalism can be of use to academia and the knowledge it produces in the spirit of public sociology.

Fundamentally, the book tries to think about the practice of journalism as a reflexive process rather than merely analysing its outputs and reporting back on the nominal efficacy of such material. Doing journalism means looking at the spaces of journalism – where journalism is made, who does it and why it happens. Questions of media labour, of audiences and of the political economy of the media sector all have a role to play in helping to illuminate what journalism is, and ultimately whether it is up to the challenge of dealing with ongoing and evolving global crisis.

Making sense of journalism

Long before he penned the idiom that Berman and countless others have adopted since, Marx started his working life as a journalist at the *Rheinische Zeitung* in Cologne and then as a correspondent for the *New York Tribune*. As Hartley (1996: 34) points out, the emergent ability of mass newspapers in the 1800s to act as sense-making practice for industrial capitalism is key to our understanding of modernity as historical process. The German Rhineland in the 1840s afforded new alliances between publics, proprietors and intellectuals at odds with the political status quo, and the decision by the Prussian crown to censor and try to shut down the paper Marx was working for was emblematic of the emergent consensus and frictions between old and new elites as well as with the general population that are still central to understanding media today. In his early journalistic work Marx began to grapple with central questions of information provision and democracy; for whom is the news written, and moreover how does it help to narrate the world around us? Marx's exit from journalism under financial pressure and political coercion will be depressingly familiar to many media workers today, who set out with high ideals only to discover that their work does not exist in a pristine vacuum.

As journalism has changed, this tension between ephemeral superficiality and the thick description of reality for shared understanding has persisted. Journalists trying to push in one direction are often pulled in another, and their work becomes part of a conveyor belt of experiences instead of a carefully curated and definitive account of reality. Hartley's description of journalism as modernity's primary sense-making practice – as its textual and organisational system – is mirrored in the narrative basis of modernity established by Benedict Anderson in *Imagined Communities*. Anderson (1983) illustrates how media and journalism were crucial to the emergence of the modern state and of the mechanisms that would facilitate the advent of contemporary concepts of the nation, and by extension of *transnational* interactions. As modernity and its systems of material exchange and thought expanded, media began to reach levels of complexity that brought in larger global flows alongside more prosaic and everyday news. Anderson (1983: 33) views journalistic narrative as fundamentally novelistic, consisting of different 'characters' of the present, be they baseball matches or the country of Mali, who leave and reappear as part of an ongoing serialisation of reality that structures our individual experiences of local and global relations. As journalists our best hope of agency is to understand and intervene in this stream of consciousness, persistently and repeatedly.

When print was joined first by radio, then by television, and ultimately by the internet, journalism co-opted and was co-opted by technologies of communication, creating new communities of interest and interaction to rival older ones. Conspiracy message boards, expat newsletters, scientific discussion forums and podcasts all reproduce this same journalistic quality of diving in and out of a montage of the present, mixing the personal and the global, and of trying to make sense of it.

Hartley's definition of journalism's textual system as part of a 'popular reality' also means coming to terms with the knowledge claims of journalism about the world at large. As the journalism scholar Karin Wahl-Jorgensen (2016: 9) suggests, 'the knowledge claims of journalism have broader ideological consequences, but are also shaped by sociological forces and prevailing power relations'. Journalism occupies a particular epistemological position in a reflexive and ongoing practice of sense-making of itself as well as of the world around it. The question of journalism's own future and the way it sees its past is thus bound up with lingering doubts about the validity of its own self-understanding.

Within this there is another thorny issue. Where is the line to be drawn between journalism and media when, as the media theorist Mark Deuze (2012) suggests, we now inhabit a 'media life' in which we do not just consume media but live in media? The distinctions between 'facts' and 'fake news' have become a rallying cry for media workers seeking to assert the moral purpose of their work over the noise of pseudojournalism and politically or commercially led content generation. Journalists and journalism researchers alike are prone to stressing the specificity of journalism as a particular practice with utopian and civic ideals, delineating and defending it against the encroachment of entertainment and social media on its tightly guarded border. Self-consciously serious political reporters in well-pressed suits whose professional ethic relies on a practised earnestness stand solemnly with their microphones in front of the institutions of state. Foreign correspondents photograph themselves in the White House press room to reinforce the unique nature of the knowledge they provide. And investigative reporters lament the decline of the red-eyed journalist sitting in the corner of the office late at night to expose the dark underbelly of power and money. Wherever we go we encounter forms of gatekeeping that constantly try to define what journalism is and is not.

Undeniably, however, journalism as a broad set of practices has carved out a normative position in the institutions of modernity, defined by 'public service, ethics, management, autonomy, membership of a professional elite, immediacy and objectivity' (Bogaerts and Carpentier 2013: 62). It is important enough to the shared understanding of what the world is and should be that its disappearance seems unthinkable, and this in part can be said to account for the desire to reassert its importance as a central institution and to so performatively guard its borders. The textual system can be rewritten, rebound and reordered, but it will always be there even if its gatekeepers are not.

Between abundance and collapse

Paradoxically for a profession which sees itself as perpetually under threat (and not without good reason), there is now more journalism than ever, or at least more of the things that claim to be journalistic. Newspapers now stay quiet on their print circulations or subscribers but vaunt the number of individual users they attract each week online. Startups draw in investment capital based on optimism about monetising potential user bases in the hundreds of millions that may or may not exist. The problems of journalism are demonstrably not down to a lack of production, nor indeed to a lack of media consumers per se, which challenges the simplistic analysis that there is no demand and no ability to supply.

Journalism suffers from a world of abundance in which, to borrow a phrase from the utopian early promises of nuclear energy, it is so plentiful that it is too cheap to meter, yet the plentiful flow of information through wire and fibre optic cables also cheapens the product. The issue with the post-scarcity journalistic society is that it is very difficult to make money from selling anything. A world where anyone can be a journalist and produce journalism at a basic level is a half-realised utopia because, as any struggling media worker will tell you, the journalists themselves do not live in a post-scarcity world where their material needs are taken care of. Entering journalism as a profession today is to enter an insecure field of precarious labour and side hustles (and more often than not overwork) to be able to produce the goods, and to acquire the cultural capital that some journalism roles still bring.

In this rhetoric of collapse and stagnation, one growth area in journalism amid the decline since the early 2000s has been the emergence of grand manifestos and much hyped innovations which claim to hold the answers to this existential crisis. Digital-first, community-driven, data-driven, ideas-driven or solutions-driven journalism and a raft of other concepts have been deployed as ways out of this existential introspection, and as a means for journalism to reclaim its role as sense-making practice and survive as an institution. The need to assert the importance of journalism in the present often manifests as a struggle for agency and professional identity, and for the continuation of organisations which regard themselves as too important to be irrelevant in the future. This is not helped by a generation of media scholars who at times have been wholly in awe of technology and the spectacle of the new toys available to them, to the point that the future has merely become a question of realising technological potential, or of the promise of deterministic technological disruption over intentional political change.

This technofetishism can confuse to deceive, ignoring the material processes which guide both our lives as citizens and our work as journalists. Journalists do not lose their jobs due only to the grand sweep of history, any more than the workers in a German car plant do because their jobs have been shifted to Asia by an invisible hand. In these grand narratives of digitisation and the entropy of the journalism industry there are competing priorities in terms of the maintenance of journalistic capacity and the priorities of profitability. This forces anyone interested in the future of journalism to confront the behaviour of capital and its role in shaping the industry. This in turn has implications for the way journalism itself conveys understandings of global economic conditions through internalisation of these narratives, and how well it can act as a medium for seeing the world at large and its myriad problems.

Building a global media

Journalism has not always immediately embraced new technology, yet technological change in both news work and audience consumption has had significant and longstanding impacts on how the news is made. When cable and satellite news with rolling updates and schedules began in the 1980s, world events became less novelistic and more like a soap opera. The advent of 24-hour cable news with the launch of CNN in 1980 marked a step change in the idea of events being reported 24/7 from a variety of locations across North America and the wider world, and as frustration with American news hegemony grew it spawned imitators in other major economies. When trying to understand the relationship between journalism and the world-spanning consequences of modernity, the development of national and regional press cultures into a globalised media functions as a shadow to the material economic processes of globalisation, while the immediacy of current affairs also increases the function of journalism as a practice of the extreme present. Most of all, 24-hour news and the internet have both contributed towards making the long sought ideal of global journalism possible, albeit in a less utopian form than many might have hoped.

Global journalism as a practice has been widely theorised and studied as an antidote to the narrow focus of national press cultures and domestic politics. With its gaze set firmly on global issues such as migration, poverty, climate change and 'bads' such as deforestation, ocean pollution and conflict (see for example Berglez 2008), the aspirations of global journalism envisage the cosmopolitan press rising to meet the challenges of globalisation and economic, environmental and informational interdependency.

Multinational media conglomerates and brands have come to dominate this global media landscape, alongside the emergence of nominally global broadcast and online organisations situated in specific territories, like Al Jazeera, which compete with established global actors such as the BBC and CNN. The BBC – as well as being a key driver of *Our World* – was a pioneer in transnational broadcasting, with its World Service emerging from the Empire Service, eventually finding form in the globalisation-ready BBC World News. This has granted it an atypically prominent position in the international media landscape for what is ultimately the public service broadcaster of a secondary global power.

What is less visible perhaps to the general public is the growth of transnational media ownership behind much of what they consume and the emergence of transnational media corporations (TMCs). Groups like News Corp with extensive assets across the English-speaking world, the French media giant Vivendi with interests in broadcast, streaming and computer games, or the Japanese holding company Nikkei, which owns the *Financial Times* in addition to a number of East Asian assets, typify the transnationalisation of ownership. These are just a few of the multiple imitators and competitors with global asset bases to emerge as asset capitalism has annexed global media. Such expansion of multinational ownership – and by extension of operations – in the media is replicated across other sectors of the global economy as financialisation and deregulation have transformed journalism and media around the world from rooted institutions to tradeable international assets. Globalisation for the media is not just a question of understanding global change as a subject, but as a process internal to the structures of journalism itself, and of how these changes have come to impact the workings of journalists as they try to make sense of the world.

Whether the transnationalisation of media has led to more genuinely transnational approaches to journalism is a point of debate. The idea of a truly global journalism exists somewhere 'between dream and reality' (Van Leuven and Berglez 2016), with the impulse to practise contemplative and comprehensive transnational journalism being held back by the inertia of domestic news cultures and local markets and audiences. As will be discussed later in relation to cosmopolitanism and the Anthropocene, the global gaze is often highly situated in the seats of editorial and corporate power. We have at once a globalised media system in which media capital is mobile and fluid, and a journalistic crisis intimately linked to this globalisation and its appetite for extraction, in which news media can even be compared to other sites of mobile extraction for profit such as oil or minerals (Hinde 2021). Multinational networks are set up to mine both the news itself and audiences as assets before moving on to new sources of income and cutting loose their unprofitable investments. Profit seeking and the short life of assets explain much about why journalism behaves the way it does, and about its enthusiasm for insecure labour and flexible work rather than long-term planning and institutional investment, but also about how it treats its audiences as transnational commodities.

It is in this context that the question of how we convey the Anthropocene as a journalistic undertaking comes to the fore, as well as that of the relationship between journalism, modernity and our understanding of the processes that affect both. As we deal with globalisation's externalities coming back to haunt us all, and with the attrition to media practice from a media market that resembles a technological oligopoly, and profit seeking from the media companies that are themselves entangled with technology companies, hedge funds and asset strippers, the future of journalism as a meaningful global practice can seem both uncertain and bleak.

Journalism in modernity's shadow

There are, of course, plenty of examples of journalism succeeding and thriving, and these pieces of good news provide the seeds of something better perhaps, but how able these survivors are to resist the dominant trends towards a reduction in journalistic services and the centrality of journalism as a resourced profession is open to debate. Seeing through this crisis is a question of utopian idealism and sober realism at the same time, but also of engaging with questions of form and practice as much as straightforward coverage of climate issues. Journalism is caught in a complex web of epistemological uncertainty, economic pressure and representational confusion.

Doing journalism in the Anthropocene means looking at where journalists will be in the near future and what they might become under the liquid conditions of contemporary media capitalism and its alternatives. It also entails looking at the modes of expression available to journalism and the technological changes which at once undermine and push forward the profession. Old divisions between technicians, editors, copy generators and writers are broken down as the prospect of several different journalistic futures appears on the horizon.

Given the vastness of the Anthropocene as an idea and the scale of the challenges presented by climate change and associated processes, it is easy to talk about these world-spanning effects and events in the abstract, to analyse what is unfolding in front of us each morning on our phones and on the TV as a spectator event. Yet all of these mediations rely on the physical work of journalism and journalists, many at the heart of these violent changes. Modernity itself may be on the table, but making sense of it is about the banal as much as the spectacular, a question of the everyday work of doing, as much as a time for grand manifestos and increasing degrees of abstraction. Ultimately journalism is work, albeit a particularly public kind that has far-reaching implications for the people who do not practise it.

There is also another realisation that journalism must contend with. Not only has journalism failed to grasp the sheer urgency of the climate crisis/catastrophe/emergency (the jury is still out on the semantics of each), by looking back at how journalism has been enmeshed with the fabric of modernity we see how it has often been complicit in the construction of the Anthropocene and its norms. Journalism fails not merely the task of addressing the problems we now face under the Anthropocene rubric, but has actively contributed to them and continues to do so. This is harder for journalists and journalism researchers to swallow than merely accepting that they do not perhaps cover Anthropocene topics like climate change and biodiversity loss as readily as they should, moving journalism as a profession from a place of plausible deniability to a place of direct culpability. The ultimate aim of doing Anthropocene journalism is to try and navigate this world and to try and move it on to something less disastrous, more stable and more solid. To answer the question of what this means for reporting the world now and in the future, we have to consider both the death of 'nature' and the death of journalism as we know it, and face with sober senses the future we have created.

2

The End of Nature and the End of Journalism

In 1989 Bill McKibben, the American environmental journalist and campaigner, published *The End of Nature*, a bestselling work of popular non-fiction discussing climate change and the totality of human impacts on the planet. At the heart of McKibben's breakout success was his epiphany that human influence on the earth's climate and ecosystem meant that our view of nature as a pristine other was no longer tenable. In the American backwoods, a new view of how we should report the world took root that would have a significant impact throughout the 1990s. Two years later the sociologist Bruno Latour published *We Have Never Been Modern* [*Nous n'avons jamais été modernes: Essai d'anthropologie symétrique*] (1991) in France, taking aim at the dichotomy between the human and the natural world and arguing for an understanding of 'nature' as intimately entangled with the social and cultural spheres. Building on Latour's previous work in actor–network theory (ANT) and the realisation that real material things both human and non-human, the living and the inanimate, exist in networks of agency with one another, Latour placed studies of modernity and the 'modern world' within processes that were regarded as external to humanity. Whereas McKibben saw the 'end of nature', Latour (2016) had in his own words 'never believed in Wilderness', growing up in a rural French landscape that was already irrevocably changed by humanity when the Romans invaded Gaul two thousand years ago.

Latour and McKibben occupy different roles and positions in the constellation of ideas that have come to form the contemporary environmental movement and in the academic debate on the future of both the planet and civilisation. Yet both became household names and veterans of debates on climate and society at universities and policy summits across the planet. Though by no means unique, *We Have Never Been Modern* and *The End of Nature* are regarded as watershed texts in the recentring of the modern environmental debate at the beginning of the 1990s and the looming spectre of climate change on the horizon.

From a journalistic standpoint, McKibben's work has assumed canonical status as a prescient text which heralded a step change in how environmental journalists and activists should address global risks and the behemoth of human impacts. Using the American landscape – once a land of endless opportunity and nature to be tamed – to argue for a re-evaluation of how we look at the natural world, McKibben was able to tap into wider currents around global environmental protection at the end of the 1980s, but also speak to a particularly American sensibility and its underlying anxieties.

McKibben's work was, by his own admission, incomplete and America-centric, written by a 27-year-old with a burgeoning but incomplete set of ideas who was fortunate to be backed up by the relatively deep pockets of the *New Yorker* magazine. Its appearance might be considered serendipitous as much as the result of its uniquely strong and pioneering argument. Working for a well-resourced publication with the means and the intellectual curiosity to support such work, McKibben's experience was in many ways atypical of the average journalist even at the beginning of the 1990s. Nonetheless, the resources available for projects of this scope would be unrealistic for many long-form journalists now, forced to scrape together time and costs in the pursuit of deep storytelling. No longer in a halcyon world in which magazines and newspapers were able to bankroll forms of journalism that brought no immediate financial return and to provide depth and reflection, these practices of reflective long-form writing have today come to be considered as 'slow' journalism, and as a means of reclaiming hallmarks of quality and depth (Le Masurier 2015). This nostalgia for the time and solidity of mass print empires today manifests as a form of radical practice in the speed of contemporary media work, but also as a lingering desire to rewild the barren intellectual landscapes of modern journalism.

It is rare in today's journalistic labour market to have the financial stability and patronage of an institution that allows such freedom, but at the start of the 1990s the internet was still the preserve of computer scientists and basement enthusiasts. News stands were a familiar sight on street corners and railway stations across Europe and the rest of the world, with a finite series of canonical titles lined up with their front pages on display – which in their limited offering also embodied an institutional stability. Today the newsprint shelves are the preserve of only a few titles and organisations, themselves often heavily rationalised, and the print magazines and newspapers compete for space with bottled water, protein shakes and iPhone accessories. As the news stand began to vanish from the fabric of everyday life, our awareness of the enormity of environmental risks has risen. Interest in climate change is higher than ever before, with record amounts of coverage, but what about nature, so long the concern of environmental journalists and their fellow-travellers in the wider environmental movement?

Nature has no problem soliciting awe and support, from the runaway hits of the BBC's *Planet Earth* to the use of idyllic flora and fauna as backdrops for expertly curated personal Instagram brands, or the charismatic animals of the internet that loop on autoplay in social media feeds. Lost dogs, interspecies friendship and the vistas of a hundred thousand mountains summited and beaches visited draw us into a fundamentally unreal, sanitised and curated form of the natural world. This packaged and awe-inspiring media spectacle trades on old modes of seeing the world in a digital zoo, but these charismatic animals and unspoiled panoramas are not so much engagement with nature as an escape from it, eschewing reality in a world of pure representation. Putting nature on screen, whether as the producer of expensive documentaries or a teenager using a smartphone to film the exploits of a dog on a front porch, is not a phenomenological encounter with the environment at large.

There is no shortage of content, but not all content is journalism and not all content helps us to make sense of situations.

The retreat of environmental reporting as classically understood from the front pages and the priorities of newsrooms has been marked since a high point of awareness in the 1980s and early 1990s, when a large number of environmental risks demanded significant news coverage in developed countries – not least pollution (and the toxicity of deindustrialisation), habitat threats, ocean acidification and the ozone layer. A groundswell of environmental feeling in the developed world manifested in the Rio Earth Summit, a United Nations (UN) conference which formed the basis of the Intergovernmental Panel on Climate Change (IPCC) and the regular UN COP (Conference of the Parties) climate summits which are so critical to our shared future today.

The 1990s thus presented a world teeming with risks, or 'bads', looming threats to wellbeing which transcended the solidity of so-called first modernity and resulted in Ulrich Beck's (1992) diagnosis of 'second modernity', where supposedly promising aspects of modernisation returned as revenants in the form of pollution and climate change, but also as the means by which the technological assemblages of modernity undermine its own apparent certainties and structures by destroying its own institutions and introducing new risks.

Environmental journalism in the 1980s and 1990s can be seen to function within this framework of risk management, in which one of the primary functions of environmental reporting was to raise awareness of issues and bring such risks to public attention. The concept of global risks being theorised at the time in academic sociology facilitated a picture of a society which was coming to terms not only with its own globality, but also with the existential threats that loomed in the vacuum of the post-communist world. Overfishing, nascent climate change, ocean pollution, saving whales, opposition to nuclear energy and indigenous rights became primary topics of environmental journalism.

Environment desks in news organisations were seen to fulfil an important watchdog role, contributing to a long established tradition and performance by the media as a democratic institution charged with holding government and business to account for its behaviour. In this ethic of journalism, relying on strong institutions and the ability of newspapers and broadcast media to be both commercial enterprises and important civic platforms, environmental journalists were able to establish themselves as beat reporters alongside more familiar profiles like politics, education, health and economics. Crucially, although environmental journalism is necessarily political, it also embodied a new kind of politics which did not strictly correspond to traditional notions of left and right. As a consequence, environmental journalism found itself welcome in the corridors of media power in a way that other radical social movements were often unable to, and by extension in the broader frameworks of 1990s liberalism which subscribed to a philosophy of depoliticised risk management. This duality continued to characterise environmental journalism throughout the 1990s and into the 2000s, with content appearing in one section of newspapers or on broadcast news which clashed heavily with what appeared elsewhere on the same platform. As an embodiment of the wider dissonance about the environment, however, this environmental coverage did reflect its publics.

The early 1990s was a good time to be an environmental journalist for both ideological and altruistic reasons. There was an upsurge of interest not just in climate change – or global warming as it was generally labelled – but in myriad issues of air and water pollution. Environmental correspondents were appointed at major broadcasters and newspapers, and the UN initiated the first attempts at global climate governance. This was aided by the fact that as well as being at the end of nature, we were also felt to be, in the hubristic words of Francis Fukuyama, at 'the end of history' (Fukuyama 1989) as liberal capitalism seemed ascendant across the globe.

The post-Cold War settlement may have reduced the risk of nuclear conflict, but it opened up a new chapter in global environmental risk. A surge in interest in environmental issues was reflected in the media's preparedness to publish environmental stories, while in formerly communist areas of Europe a newly liberated media was able to report freely on long-term issues that had previously been denied or underplayed. As the millennium approached, though, the first signs of attrition to environmental journalism capacity began to be felt, and by the middle of the 2000s the media worldwide had begun to experience significant layoffs and uncertainty well beyond environmental desks.

In the US, newsroom capacity dropped by around 45 per cent in just a decade in the early 2000s, with the majority of members of the Society of Environmental Journalists since working as freelancers (Sachsman and Myer Valenti 2020: 2). From a moment of promise and a central position as watchdog for the world's problems, environmentalism as a media force was left struggling for its own survival. *The Ecologist*, founded in the 1970s in the UK and an important point of reference for Britain's nascent green political movement, stopped printing in 2009, migrating to an online site which continued to struggle financially. It finally merged with the more spiritually inclined *Resurgence*, founded in 1966 and emblematic of the crossover between environmental and radical spiritual movements of the baby boomers. Although we have to be careful how we generalise about worldwide trends based on North American and European experiences, there has been a similar broad decline and general changes to the viability of environmental journalism in other markets too. As environmental reporting is not a core product in the offering of most outlets, it has fallen by the wayside and has been at the forefront of cuts to specialist correspondents and editorial teams in the drift towards more generalised content producers and flexible newsroom labour.

Environmental journalism is also a loaded term; where it starts and ends is uncertain, and its professional ethic and identity – like academics, who are often both expert and activist at the same time – is intimately bound up with the types of people who do it. Their personal histories and their modes of being and doing underpin the environmental movement as a whole and help to shape it, occupying a peculiar and specific role as movement intellectuals alongside the straightforward work of reporting the news. When we talk about the environmental beat we are also talking about a flexible concept, from the business of landscape conservation and 'protecting nature' to the intricacies of urban design, transport policy and the entire global economy. How and where environmental journalism first came into being depends on what is considered to be environmental, but also on who is writing the history.

The genesis of modern environmental journalism is often traced to the American biologist Rachel Carson's (1962) *Silent Spring*, a process that was aided by the book's appearance in English, but also by its coverage in elite media and its status as a relative bestseller (Parks 2017a). In this neat narrative of modern environmentalism, Carson's work heralded a conceptual shift in understanding the environment as a complex system which was more vulnerable to human impacts than the immediately visible signs of destruction such as logging or meddling with the flow of rivers. Carson's talent was marrying well-researched scientific arguments with a developed narrative style, tapping into existential worry beneath the surface of postwar consumerism and corporate industrial expansion. Her book was notable for its identification of the destruction of the environment through use of deeply embedded industrial processes, and she invoked hard science to take on the power of American pharmaceutical and agricultural giants in ways that fundamentally damaged their standing in US society. Had Carson herself not died soon after the book's publication she would no doubt have become a leading figure of the environmental movement into the late 1960s and 1970s, but her death did nothing to diminish the popularity of *Silent Spring* among the generation of environmental activists and writers who consumed it.

Silent Spring's position as a foundational text of the modern environmental canon is not in doubt, but a book written in English by an American and disseminated worldwide reveals the hierarchy of global knowledge production that remains in place today. The United States – at the heart of the so-called Great Acceleration in capital, population and human impacts after 1945 – has also produced some of its most well-known critics. The environmental movement and its journalistic footprint is dependent on the broader cultural flows of media and modernity, its networks of communication and its lingering malcontents.

Those malcontents had long sought a magic bullet to penetrate the consciousness of the public and persuade them of the truths that were self-evident to the activists and scientists of the environmental movement from an early stage. A decade after the declaration of the end of nature by McKibben some of them thought they had been gifted that magic bullet. In the year 2000 at a conference in Mexico the Dutch atmospheric chemist Paul Crutzen – who also discovered the hole in the ozone layer and received a Nobel Prize for his work – stood up and objected to what was being said at the front of the room. The cumulative impacts of modernity had removed humanity from the established geological and atmospheric norm of the Holocene and created a new epoch, he said. He called this age the Anthropocene.

Welcome to the Anthropocene

The Anthropocene concept was clarified by Crutzen and the ecologist Eugene Stoermer in an academic newsletter that same year (Crutzen and Stoemer 2000). The Anthropocene, so the story goes, changes everything. By recentring environment and modernity as part of the same large process, its proponents argued, it brings together the disparate impacts of human growth across the planet in one all-encompassing concept, and moreover a concept which is not rooted in the typology of social science or the periodisation of academic history but in the clear and unambiguous material markers provided by the earth sciences.

Ironically for a term invented by earth scientists, the International Union of Geological Sciences (IUGS) would subsequently reject the official classification of the Anthropocene as a new geological epoch after 15 years of discussion, yet in those 15 years the Anthropocene became a central framework for understanding the global future and lives on in the social sciences, humanities and earth sciences as a foundational idea.

The IUGS also has some fellow-travellers in its rejection of the Anthropocene concept, though for different reasons. While the geological definition of the Anthropocene has been contested on the grounds of practice and the burden of proof, analytically it has been challenged by voices as diverse as the Marxist historian Andreas Malm (Malm and Hornborg 2014) or the earth scientist Erle Ellis, who has argued that although Anthropocene conditions are very real, there is in fact a 'Deep Anthropocene' that stretches back further than the recent events of anthropogenic climate change, nuclear testing and resource extraction that mark our present (Stephens, Ellis and Fuller 2020). Similarly, where the Anthropocene is accepted in principle, it has been unmade and reworked with enthusiasm by a plethora of human geographers, sociologists and popular science writers. The visual culture scholar and art historian T.J. Demos (2017: 22) suggested that the idea 'anesthetizes politics' by imagining a singular globality in the benign language of globalisation, and the geographer Kathryn Yusoff (2018) among many others has suggested that no account of the idea is complete without understanding the history of geology itself and the racialised implications of Anthropocene thinking. Beyond the world of climate deniers and contrarians, however, the facts of the Anthropocene idea are not disputed. Instead, as social scientists and natural scientists alike are prone to doing, the term is subject to a seemingly unending quest for conceptual unification and epistemological purity that would put the doctrinal committees of the Catholic Church to shame. If nothing else, it shows the precariously social nature of scientific knowledge both in and out of the lab, and indeed the concept's inherent mediality and its contestation is what makes it so interesting for those of us interested in public life and how knowledge travels through it.

For journalism the Anthropocene promotes and provokes a series of questions. Firstly, how has journalism chosen to portray the Anthropocene to date, and conversely how does the Anthropocene problematic affect how journalism understands itself and its subjects? The answers depend somewhat on how the Anthropocene is defined, and also how it has come to be used by a range of different voices in the public media and wider society. It also comes complete with multiple competing genesis stories, beyond the arrival of a chemist in a Mexican conference hall, that influence how it should be understood, what values are attached to it and whether or not we can quickly and easily get out of it.

Among the broader debates, the consensus is that the Anthropocene might be said to begin at one of two points in (relatively) recent human history: 1610, which showed the greatest dip in atmospheric CO_2 levels before an exponential rise to what we see today (Lewis and Maslin 2015), and 1964 with what is seen to be the high point of radioactive fallout in the global ecosystem (assuming no new large-scale nuclear disasters or nuclear conflicts in the future). These two dates, based on analysis of material markers, feature two distant but related aspects of the broader Anthropocene narrative, namely the occupation of the Americas by Europeans and the environmental changes that followed, and the postwar Great Acceleration which gave rise to the familiar hockey stick graphs of carbon emissions, GDP, telecommunications and a host of other markers of humanity's exponential impact. This is made more complex by the fact that, in between, both the invention of the steam engine and the cascading of coal power across the globe began large-scale fossil emissions into the atmosphere and kick-started extractive industrial modernity in Western Europe.

Industrial modernity has unleashed not just climate change, but distinct physical transformations to the earth itself. There is now more synthetic material by weight on earth than biomass (see Elhacham et al 2020) – that is to say, human-made objects outstrip the weight of all organic living things including the entirety of the Amazon rainforest, the boreal forests of Scandinavia, Canada and Russia, and the jungles of sub-Saharan Africa and Asia. This seems incomprehensible, but the 20th and 21st centuries have been an escalating love affair with concrete, a material which is not only carbon-intensive but hugely heavy. Concrete has literally been the foundation of the Great Acceleration that took hold after 1945; without concrete there are no skyscrapers, motorways, pop-up cities or other megastructures.

The late Australian chemist Will Steffen and his co-researchers have argued that it is only during the Great Acceleration after the Second World War that earth system changes become markedly different enough from the envelope of Holocene conditions to merit the label Anthropocene (Steffen et al 2015), even though the processes which eventually came to drive these changes such as the steam engine (and latterly the internal combustion engine) began far earlier. Here it is the Anthropocene's exponentiality – modernity stronger and faster – that becomes its defining feeling. This is also where we begin to see overlaps with media and journalism through the lens of modernity and the explosion in mobility and communications throughout the 20th century.

Implicit in this is the question of how we understand the Anthropocene as a sociopolitical phenomenon. The Anthropocene did not just happen, but was the result of specific social and economic processes which continue today and will do so into the future. Several leading Anthropocene scholars, such as Donna Haraway (2015) and the economic historian Jason Moore (2017), have suggested that the Anthropocene should in actual fact be called the Capitalocene. According to them, the degradation and altering of global systems humanity has caused is fundamentally a story of capitalism's spread across the planet with the help of colonial European powers and, latterly, the hypercharged economies of the United States and China with their networks of global labour and extraction. The situation we now find ourselves in is the result of political and technological processes which have fundamentally been about the pursuit of capital accumulation, economic growth and the acquisition and transformation of natural resources into commodities.

This is perhaps best seen in the 22 Anthropocene markers suggested by Will Steffen and his colleagues (Steffen et al 2015). These metrics and the resultant Great Acceleration graphs stemmed from a project of the International Geosphere–Biosphere Programme (IGBP) in the early 2000s, trying as far as possible to accurately map the historical growth of multiple processes from 1750 to the first decade of the 2000s, and to show not only the footprint of the Anthropocene, but how these marker had evolved in tandem with social and technological aspects of modernity in a broadly exponential way.

The data produced by Steffen's team identified two basic types of processes: socio-economic trends and earth system trends. The measures used are shown in Tables 2.1 and 2.2.

In the familiar semiotic of the Anthropocene, many of these earth system trends look very much like the hockey stick graph of carbon emissions which has become a ubiquitous sight as each year brings new rounds of climate negotiation. Others are somewhat more restrained, but all show the same broad tendencies. In the decades since 1950 all of these processes have accelerated to some degree, most of them exponentially. All of them leave us with cause for concern, but metrics such as transportation, telecommunications and urban population also reveal how entangled the detrimental impacts are with modernity as we know it, and indeed with the potential solutions.

The changes to the climate, destruction of habitats and the general degradation of the biosphere are not unfortunate mistakes or oversights of modernity, but integral to it. On the graphs used to map Anthropocene processes, the more familiar shapes of growth in atmospheric carbon dioxide, phosphates and intensive farming are mirrored by an almost identical line plotting the growth of communications technology. The Anthropocene is not just a process of environmental change, but also the exponential growth and transformation of the technological assemblage which forms the basis of media practice. Printing presses, railways, telephones and urbanisation all contributed to the development of modern mass journalism, but also to the information society and the distancing and disembodying that it entails. Journalism and its history are intimately entangled with this wider story of human acceleration. Media is not just a way of seeing these Anthropocene processes but an active participant in it.

Table 2.1: Socio-economic trends of the Great Acceleration

Indicator	Trend description
Population	Steady growth from 1750, accelerating sharply post-1950, reflecting rapid global population increase.
Real GDP	Minimal growth until 1950, followed by a steep increase, showing economic expansion during industrialisation and globalisation.
Foreign direct investment	Negligible until the mid-20th century, then sharp growth in the post-1950 period, indicating globalisation.
Urban population	Gradual rise before 1950, then accelerating sharply due to urbanisation and industrial growth.
Primary energy use	Slow increase pre-1950, followed by rapid growth, linked to industrial energy demands.
Fertiliser consumption	Almost flat pre-1950, then exponential growth as agriculture became industrialised.
Large dams	Minimal pre-1950, then substantial increase due to infrastructural developments post-1950.
Water use	Gradual increase until 1950, followed by rapid growth due to industrial and agricultural expansion.
Paper production	Steady increase, with significant acceleration post-1950 due to industrialisation and education demands.
Transportation	Negligible growth pre-1950, then sharp rise with the proliferation of motor vehicles and global transport.
Telecommunications	Minimal growth pre-20th century, then exponential increase post-1950 due to technological advancements.
International tourism	Very slow growth until mid-20th century, then exponential rise with better global connectivity.

Source: Adapted from Steffen et al 2015

In a world of scientific omniscience, the latter half of the 20th century might have looked quite different; discussion of the risks created by technology and various energy sources was certainly widespread, but the Anthropocene perspective – in which humanity could have genuinely world changing impacts – remained far from the mainstream. The idea that nature itself could become corrupted by humanity to the point of system failure was unimaginable to many, and this has remained a useful and functional argument for climate denialists and their funders. Furthermore, even when Anthropocene processes were accepted as real, such visualisations of future payback for the debt we have accrued were limited to the work of science fiction writers and Malthusian environmental activists operating at the fringes of the debate. It is also worth noting that now, when we are more

Table 2.2: Earth system trends of the Great Acceleration

Indicator	Trend description
Carbon dioxide	Gradual increase until 1950, then sharp rise reflecting industrial emissions and fossil fuel use.
Nitrous oxide	Slow growth pre-1950, but rapid acceleration post-1950 linked to agricultural practices and industrial activities.
Methane	Gradual increase pre-1950, then significant growth due to agriculture, energy production and waste management.
Stratospheric ozone	Sharp decline starting in the late 20th century due to chlorofluorocarbons (CFCs), with some recovery trends.
Surface temperature	Relatively stable pre-20th century, then steady rise, accelerating post-1950 due to global warming.
Ocean acidification	Minimal change until the industrial era, followed by rapid acidification linked to increased CO_2 absorption.
Marine fish capture	Slow increase until 1950, then exponential growth due to industrial fishing, levelling off in recent years.
Shrimp aquaculture	Negligible pre-1950, then exponential growth post-1950 reflecting increased demand for seafood.
Nitrogen to coastal zone	Minimal change pre-1950, followed by rapid rise linked to agricultural runoff and industrial pollution.
Tropical forest loss	Gradual loss pre-1950, accelerating sharply post-1950 due to deforestation for agriculture and logging.
Domesticated land	Slow increase pre-1950, then significant expansion due to industrial-scale agriculture and urbanisation.
Terrestrial biosphere degradation	Minimal degradation pre-20th century, then sharp increase post-1950 due to habitat loss and human activities.

Source: Adapted from Steffen et al 2015

keenly aware of the future and what awaits us than ever before, we still show few signs of actively seeking to stop or transform the systems of capital and economy that are doing us long-term harm.

On the very pragmatic level of dealing with its consequences it does not matter when the Anthropocene began, only that we know that we are in it and that being in it is not a good place to be for any of us. Pacific Islanders with receding shorelines are probably unaware of and untouched by the attempts to clarify what the Anthropocene is in the seminar rooms of New England and old England, but we all deal with the consequences. Because we are in it, it impacts on every aspect of our lives, well beyond the classical concerns of environmental journalism. The Anthropocene is the air particulates you unknowingly breathe from the cars driving through

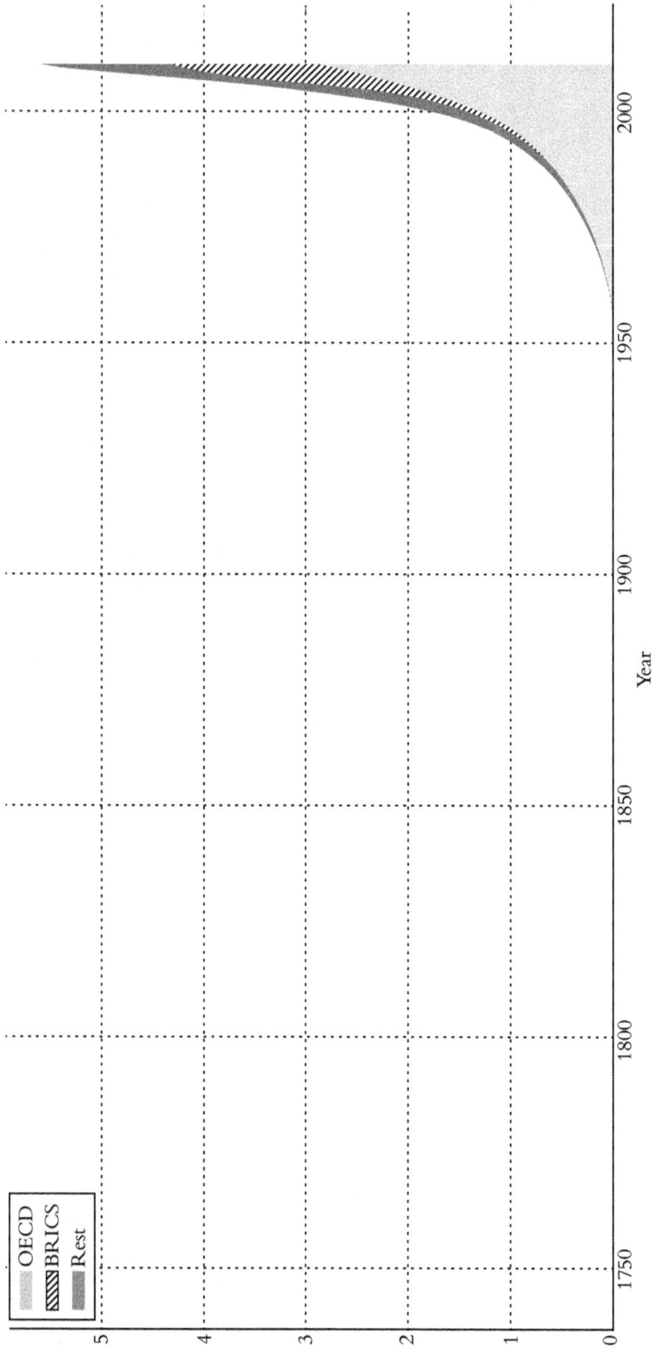

Figure 2.1: Visualisation of the exponential growth of telecommunications in the Great Acceleration using data from the Anthropocene markers project

your neighbourhood, the flooding caused by a new housing development in another town 10 km away upriver, the zoonotic coronavirus that led to an economic crisis in which you lost your job, the cancer caused by endocrine disruptors that kills you in your eighties after modern medicine has prolonged your life or, as Rachel Carson (1962) predicted in *Silent Spring*, the absence of birds in the woods.[1]

We cannot perhaps afford to split hairs over exactly what the Anthropocene is or is not when its impacts are manifested everywhere with potentially exponential frequency. The only certainty to the Anthropocene is that we and our descendants have to begin the road out of it.

This intergenerational acknowledgement is critical; we cannot merely choose to exit modernity and the Anthropocene in the space of a few years. Modernity is not a choice or a consensual democratic project, and as Ulrich Beck glibly noted, stopping it through some kind of ethic of technological restraint would be like using a bicycle brake on an intercontinental jet (Beck 1992: 106). Similarly, the Anthropocene is us and we are it. The writer and ecocritic Roy Scranton has summarised the bind between cause, crisis and resolution by noting carbon's role in shaping our collective forms of social life and creating a single integrated economy. Without the energy and information structures sustained by carbon there would not be any global civilisation to save (Scranton 2015: 53).

The invention and end of wild nature

The woods of the north-eastern US that form the backdrop to both Bill McKibben's and Rachel Carson's work had already undergone a great transformation through the purging of indigenous peoples and wildlife upon European colonisation. The pristine forests beloved of the American environmental movement form part of a combined continent that some have suggested was home to up to 60 million people before European colonists arrived (Koch et al 2019). Like the 'wild' Scottish Highlands or the wilderness of northern Scandinavia and Russia, these places had long human histories before the conservation movement set eyes on them. As we unpick the tragedy of the Anthropocene we also have to ask harder questions about where we have chosen to find that tragedy and tell stories about it.

In environmental journalism, and indeed in a lot of scholarship from environmental communication, we can discern a perpetual hunt for the macro-narrative or the magic bullet. The significant body of work which has emerged in the past two decades on climate communication in particular seeks to provide a grounding to these ideas, yet implicit in much of this thinking is the idea that information provision, democratic agitation and instrumental communication are enough on their own. The growth of climate communications on the one hand and the enduring popularity of nature on the other speak to a desire for ideological hegemony on the part of deep green environmentalists and liberal reformers alike, both of whom often ignore the depth and complexity of the problems they are seeking to solve and their social context.

The communication of this ethic, and of environmental journalism and nature writing as a form of ideological epiphany, is seen clearly in Bill McKibben's description of his own realisation of the end of nature in the backwoods of New York:

> Almost in a flash, what I considered the deepest meaning of this news came to me. It was the insight at the heart of *The End of Nature* – that for the first time, human beings were managing to alter and degrade everything around us, that our impact on the environment, which before had ended at the edge of our villages or the margin of our fields, was now ubiquitous. (McKibben 2005: 183)

This moment of realisation deep in the woods of New York state is a familiar trope in the world of environmental writing, in which maverick journalists are called to persuade others of a truth that has been revealed to them.

American environmental historian William Cronon (1996: 70) has described the preoccupation with wilderness and modernity as a 'creation of the culture that holds it dear, a product of the very history it seeks to deny'. Cronon argues strongly that the creation of the American – like the Scottish – wilderness is a projection of a modernity constantly uncomfortable with its own foundations, and that the idea of wilderness occupies a specific place in the expansionist instincts of high modernity. By extension, those who wander into the wild in search of deeper truths do so still trapped in the spiritual scaffolding of the 19th and 20th centuries.

The Scottish critic Kathleen Jamie (2008) has criticised the practice of a 'lone enraptured male' who imparts deeper truths to the public and expects them to assume the enlightened world views of upper-class university educated men in a manifestation of their cultural power. Liberal ideas of popular consensus and information provision can obscure the networks of ideology which dictate how we treat the environment, and grant citizenry more agency than is often the case, combined with a belief that the knowledge bases of elite environmentalism are the points around which others should coalesce. The sincere belief of many that environmental problems could be tackled through moral enlightenment and informed discussion alone appear increasingly naive as liberal institutions continue to act out of line with the available science, all while the information deficit model of environmental communication crucial to understandings of policy response in liberal democracy has been widely questioned (see for example Moser and Dilling 2011). It is not because we do not know enough about climate change, or soil nitrification, or anything else that threatens our collective future, but because we lack the capacity to turn that knowledge into power and to stitch it into our collective understanding of the world to come. Climate change is not a mystery, but unpicking the world system is.

In Britain, where understanding of the environment is often dominated by pastoral constructions of how the landscape should be, nature journalism is big business and the British Isles has a hugely active nature lobby across the political spectrum. Media figures such as the Cambridge fellow Robert Macfarlane have forged careers off the back of this stylisation, but also the desire for wilderness on a densely populated island and its transformation into a product that can be sold as a metropolitan daydream. As Jamie (2008) suggests, 'it's only recently that we, with our (almost) guaranteed food supplies, motor engines, vaccines and antibiotics, have begun to make our peace with these wild places, and to seek recreation in land which was once out to kill us'. Just as has happened in the US conservation movement's fascination with preserving America's 'wilderness', Britain has reinvented the 'wild' as a cultural object to counter the brutality and ugliness of contemporary life and politics, and to covet a potential escape from the modernity it so forcefully helped to create.

The popularity of such perspectives in the presentation of the environment and elite environmentalism is linked to the relative hegemony of the upper classes in media industries. In the United Kingdom, for example, data shows that over 50 per cent of journalists were privately educated, rising to over 80 per cent of editors (Sutton Trust 2019). Journalism's class problem is entangled with its professional and epistemological practice and – sadly for any young ambitious journalism students – journalism in many countries is not a meritocracy but rather a question of cultural capital and staying power. The complex gatekeeping of the journalism industry means that it is still the privilege of the few to report generally, with Nikki Usher (2021) showing how in North America the decline of journalism and the concentration of power in metropolitan centres has made it even more of an elite profession than previously. It is through this lens that we have to look at the growth and decline of environmental journalism in the Global North, replicating and reinforcing existing social hierarchies even when it has altruistically sought to aid awareness of environmental problems.

In the United Kingdom the cult of the new nature writing in which the environment becomes a canvas for metropolitan longing and the journalism which accompanies it has been significant. Both left-wing and conservative newspapers have heavily narrativised crazes such as wild swimming, and regularly publish lifestyle pieces about nature's healing and restorative power. In Scotland conservation groups and wealthy environmentalists actively buy up and cultivate rural land in the name of rescuing it, despite that land historically having been populated or supporting existing communities.

Scotland now has huge volumes of 'wild' land, and its cultivation and protection is a £multimillion industry with government backing and investment from international capital, yet its cities remain congested, with inadequate housing and poor public transport, its urban green spaces gobbled up for development. To see the problem with taking nature as a guide for tackling Anthropocene problems you need only stand on the wild hills above Glasgow and look down at the maze of motorways in the deprived communities of the country's Central Belt. There can be no more escape to the wild.

Journalism after nature

As with the rest of the journalism industry, there is now more environmental content being put in front of more eyes than ever before. Environmental journalism has gone from being an elite enterprise with a relatively small readership to becoming a general feature of the news-scape, yet the number of environmental journalists operating full time is a fraction of what it once was. The decline in permanent employment in the newspaper industry is merely one factor in this picture – it is also possible to trace a generational shift between the radical ecologists of the 1980s and 1990s, themselves products of the countercultural environmentalism of the 1960s and 1970s, and a new generation of media workers in which awareness of climate change and other bads is tied up with a more general experience of precarity and social liquidity.

The difference between Anthropocene journalism and environmental journalism is to be found in its totality. Whereas environmental journalism has generally focused on the health of the natural world, Anthropocene journalism has a different gaze and takes a wider interest in modernity and significant aspects of human society at a macro level beyond. The squeeze on environmental reporters has led to new creativity in gaining support and exposure for climate stories from media gatekeepers (Gibson et al 2016), and those wishing to develop mediation and dissemination of environmental knowledge in new directions feel that conventional styles of reporting may well be redundant anyway. Veteran former *New York Times* reporter Andrew Revkin summarised the challenge facing veteran reporters on the green beat with the rhetorical question, 'Do I want to spend the next 20 years of my life writing good conventional journalism about global warming when I know it's not effective?' (Sobel Fitts 2014).

Rather than being concerned with individual threats or environmental problems to be solved, Anthropocene journalism necessarily attempts to interrogate the broader systems of human society and their relationships to earth systems as a process of technological, social and ecological entanglement. Climate change is obviously one of the most important components, but as the Anthropocene graphs developed by Will Steffen and his colleagues show, it is far from being the only factor to consider. The Anthropocene is not just the trees under threat from a local housing scheme or pollution in the local river, but the speed of everyday life, the clothes made from plastics derived from oil, and the flight booking website for the intercontinental holiday that forms part of the infrastructure of Anthropocene modernity.

In congress halls around the world politicians and business leaders meet to discuss climate change and the collapse of the media in much the same way. In both questions futurology has become a lucrative undertaking, with an array of speakers at multiple conferences declaring that the future will be different, before zeroing in on a particular technology or innovation, pilot study or example which they claim will fix the existential threat. This futurological tendency, combined with the internalisation of the idea that journalism is 'the first rough draft of history' as the former *Washington Post* editor Phil Graham supposedly claimed, invites us to imagine that the here and now is the defining moment.

That the *Washington Post* is now owned by one of the world's wealthiest people, Amazon's Jeff Bezos, is also an irony that shouldn't be lost on historians of journalism and media. Bezos is critical to one particular view of the Anthropocene which presents 'the future as an opening up of the relationship between the Earth, technology, and humanity; a cosmopolitan movement from which all will benefit and which will guarantee the future of the planet and of the entire human species' (Abbinnett 2019). This blue skies approach to the future of the media and all of its readers and nonreaders relies on a manifest belief that great men or their inventions will be the giants of the Anthropocene, and that journalism can be fixed through innovation and market capture in the same way that ecomodernist accelerationists tout currently nonexistent technologies as the solutions to currently existing and increasingly aggressive problems.

Roundtables, keynotes from industry leaders and stakeholders, and a discursive framing of the future as something which can be negotiated by assembling people in conference halls to relay the results of particular studies and data sets allow us to project certainty in spite of the precarious structural situation. This is the context in which environmental journalism now operates, grappling with an increasingly insecure foundation and trying to make sense of a world in which multiple environmentally linked issues loom large, seeking meaning and operability in the economic and technological assemblage of digital capitalism. If nature is indeed dead, then so perhaps is nature journalism as anything more than pastoral escapism. What follows is uncertain, but the journalism of the Anthropocene cannot continue to look the same, for both intellectual and organisational reasons. The death of traditional environmental journalism, including its anti-modern tendencies and its relatively elite make-up, is perhaps not to be mourned.

Journalists like to cast themselves alternately as truth tellers, intellectuals, arbiters of public morality and agents of change. Ironically given this self-awareness of journalism's role in society, they also exhibit a lack of reflectiveness about their own position as functionaries, technocrats and ultimately as labour too. The conflict that haunts environmental journalism as much as other branches of the sector is the performed ethic of excellence and purpose set against the reality of the material constraints of ownership, resourcing and readership.

As we grapple with the end of nature as a single object and of journalism as we have understood it, new modes of agency are needed, as are hard questions about environmental journalism's own self-image. Ultimately that means seeing the Anthropocene and its implications not just as a new label for old problems, but as an invitation to reimagine how we see the broader structures of modernity itself.

3

The Anthropocene in Media

How the Anthropocene as a name has been used and abused by journalists is important in understanding what it might be useful for, but equally what it is not able to do. At times it has been employed merely as a rhetorical flourish to lend journalism a particular temporal heft, the ultimate macroconcept possessing both novelty and scale, but also called upon as a catch-all term for all that is bad in the world. Liberal interpretations have used it to reinvent classical environmental concerns as well as to cast contemporary political pressures as part of some kind of new emergent system. Divorced from its scientific basis, for some the Anthropocene began in around 2010 and was useful for as long as its dawning horror could map closely onto contemporary politics, aided by the quiet nod of recognition that comes from membership of the Anthropocene readers' club.

For many of its adherents it has merely been analogous to globalisation, reheating some of the concerns of the 1990s for the long 20th century. It has been used as a thematic basis for experimental electronica by Brian Eno and as a marketing tool by the luxury watchmaker Christopher Ward, as a topic for operas, and most importantly in countless academic funding applications sensitive to the popularity of the term in broader media discussion. The need for novelty and the ability of the Anthropocene to provide it meant that the science of the Anthropocene and the results of the working groups determining whether it was a geologically valid concept have been secondary to the appetite for new forms of old phenomena. Instead of exploring nature there are now a whole host of books in which the authors explore the weird future or embark on last-chance expeditions to visit places that are melting away or soon to be destroyed forever. This is the embodiment of both excitement and loss, of wanting to be at the centre of the historical moment while being aware of the implications of that moment for the rest of humanity. At its worst this trend speaks to journalism's most narcissistic and colonial tendencies, reinventing old genres of exploration in contemporary forms to bring strange new worlds to the breakfast tables and lounge chairs of middle-class suburbia alongside their morning flat whites and granola.

The Anthropocene has also often been cast as the archvillain in the stories of global geopolitics. During the first presidency of Donald Trump, *The Atlantic* published an article headed 'Donald Trump is the first demagogue of the Anthropocene' (Meyer 2016). This value-laden invocation of Trump as the embodiment of the antidemocratic extractive client capitalism that the article saw as a feature of the Anthropocene is not unique. It is convenient to think of environmental crisis as being the result of a bad world ruled by bad people, but though contemporary modernity may be a little blunter and uglier in the United States than in Sweden or New Zealand, it is not solely the result of US global hegemony. Similarly, *The Economist*'s (intentionally) bombastic cover story with the headline 'Welcome to the Anthropocene' (*The Economist* 2011) announced the Anthropocene in the language of concerned international paternalism, a challenge for its readers and the international managerial class to take into account alongside the other stock characters of the looming future. The image of the earth as a machine that the magazine used to illustrate its cover story has become infamous as a milestone in the mainstreaming of the Anthropocene debate, showing that policy makers and economic strategists, just as well as English literature scholars and geologists, could take the term and run with it in a way that suited them.

The Economist's treatment of the Anthropocene is also indicative of a wider tendency. Where journalism has chosen to engage meaningfully with the Anthropocene as a specific concept, it has tended to be focused in elite quality publications in the Global North. The Anthropocene may be world-spanning, but its future media fossils may be limited to a relatively small patch of ground. The sociologist Leslie Sklair undertook to survey the incidence of the Anthropocene in news outlets from the mid-2000s through to 2017 (Sklair 2020) and found the existence of three main frames for the Anthropocene as an idea in the public media. The first of these was a neutral reporting tone in which the Anthropocene was mentioned in passing as part of a relatively disengaged overview. The second frame was the idea of the good Anthropocene, with its challenges leading to innovation and solutions emerging from human endeavour, and the existential threat of Anthropocene conditions creating a stepping stone to human progress. The third was a more morbid frame of pessimistic radical change, by design or by necessity. These three frames encapsulate the different imaginaries of Anthropocene thinking, not least a duality which allows the Anthropocene to exist as both an intellectual and scientific talking point and as a headline-grabbing prophecy of apocalypse at the same time.

The problem with searching for the Anthropocene in media, however, is that the cognitive framework of the Anthropocene has existed for longer than the popularisation of the term, and many of its features are documented and remarked upon in the media without mentioning the Anthropocene by name. The question is not whether the Anthropocene exists, but whether it is a prognosis or a diagnosis, and how useful it is to practitioners trying to come up with treatments for our contemporary condition.

The idea of a journalism of the Anthropocene as a brand new project is in some ways counterintuitive too. All modern journalism, tied up as it is with the spread of extractive modernity across the earth and the informational exchanges therein, has taken place in the Anthropocene. The material effects of the Anthropocene existed in media long before it was given a name and attributed specific characteristics. When we go looking for the Anthropocene we do not need to look very far, and we can look backwards as well as forwards. There is no shortage of Anthropocene subjects – the motoring pages or the business section of a news site are Anthropocene journalisms as much as the environmental pages. They merely approach it from different angles. As long as we live in modernity the Anthropocene is right there waiting for us.

Anthropocene archetypes

Given the spectacular scale of the Anthropocene, journalists are often (and for understandable reasons) drawn to certain archetypes that fit the straightforward narrative of a modernity gone awry. In terms of the size of the Anthropocene and its novelty there are a host of set subjects and forms available which can be reproduced and copied to convey the strangeness of the Anthropocene condition, particularly its overwhelming power. Trips to Antarctica, to Chernobyl, to the Brazilian interior or to rare earth metal mines in China all feed into the sense of awe of the Anthropocene as an overwhelming experience of size, scale and distance, but also of foreignness. These locations share a remoteness and a mystery which makes them a favourite of news features and television packages. In these existing narratives we tell about humanity's legacy in 'nature' there are familiar lessons and aesthetics, not least around Chernobyl, the Anthropocene disaster *par excellence*.

The story of Chernobyl is well known; in April 1986 the Chernobyl nuclear facility in Ukraine, then part of the Soviet Union, suffered a meltdown and irradiated not only the immediate area but large tracts of central and western Europe to varying degrees. Initially denied by the government, what happened at Chernobyl became a talking point around the world as radiation levels in the atmosphere spiked, and it slowly became clear that a significant nuclear disaster had taken place. The reporting of what happened next is canonised, and indeed has been successfully dramatised in multiple TV adaptations, so that the site itself has taken on an iconic position in the understanding of the Anthropocene and in our notions of post-human recovery. The decision to set up an exclusion zone around the Chernobyl plant and Pripyat, the workers' city adjacent to it, have created an Anthropocene ghost that became a regular destination for journalists and tourists alike. For others it functions as a piece in the jigsaw of Soviet memory, and more than once parallels have been drawn to Andrei Tarkovsky's *Stalker* (1979) in which curious travellers enter a forbidden zone containing unknown dangers and strange mutations. The modern Chernobyl is part existential odyssey, part Soviet kitsch.

The 'rewilding' of the Pripyat area following the departure of the human population provides both a glimpse of a post-human world and a cautionary tale about the perils of technology and its catastrophic consequences. The linear model of the Chernobyl story is a straightforward one and has been cemented in loosely agreed broad narrative. A terrible disaster occurred, made worse by the excesses of Soviet information control and public denial, but following the end of the Cold War a series of responsible international actors secured the site and today the exclusion zone is a curiosity which has to all intents and purposes been turned into a haven for wildlife and a post-human refuge. Like the Korean Demilitarised Zone between North and South Korea or the United Nations buffer zone that cuts across the island of Cyprus, it forms a vision of a future in which nature has returned and re-established itself.

Similar to what has been described as an 'anthropause' as a result of the global COVID-19 pandemic (Searle et al 2021), the Chernobyl scenario is a laboratory for removing humanity to the extent that 'nature' rebounds in new and novel ways. Though there is ample evidence to show that removing humans does have dramatic impacts on biodiversity and wildlife, in effect bubbles on the surface of the Anthropocene sphere, the reality of Chernobyl is more than a warning from history or an example of post-human utopia. Chernobyl has a long tail that continues to this day and its effects are deeply entangled with contemporary politics and notions of risk and technology.

The Belarusian journalist Svetlana Alexievich sought specifically to tackle this forgetting by looking at the ways in which Chernobyl has left human and natural scars down the years. In *Chernobyl Prayer*, Alexievich (2016) assembles the stories of those who were involved in the immediate events of the Chernobyl disaster and traces the impacts on their lives, returning to show that their injuries and experiences were not resolved neatly according to the orthodoxy of the Chernobyl story, and that the Chernobyl nuclear event produced complex new forms of human entanglement that reach into contemporary politics and deep time alike. As if to prove the point, in 2022 the Russian invasion of Ukraine saw Russian forces seize Chernobyl and threaten to unleash its toxic legacy and reopen old wounds as the site became a literal battleground for the future. The supposedly settled legacy was in fact anything but.

This core idea – that Chernobyl is not over however much the news cycle attempts to historicise it – has been developed by the Massachusetts

Institute of Technology (MIT) science and technology studies scholar Kate Brown (2020). Brown's fieldwork on Chernobyl has sought to recentre the Chernobyl disaster not as a warning from history but as a current and ongoing event. Straddling the border between historical ethnography and journalism, Brown undertook extensive fieldwork around the site, including with berry pickers in contemporary Ukraine who were collecting irradiated berries and mixing them with clean berries for export so as not to violate official limits, and exporting them to the European Union (EU). Chernobyl has not gone away because its reactor has been capped and sealed.

Chernobyl is a fitting case study when looking at the stories media tells about the Anthropocene, because it is so well known and yet simultaneously hardly known at all. Fears that the 2011 nuclear accident at the Fukushima power plant in Japan should become 'another Chernobyl' illustrate the way in which it has become a baseline for what can go wrong. Yet, as Brown goes to great pains to point out in her studies of the Ukrainian experience, Chernobyl was not an anomaly but a point on a broader and quite intentional irradiation of the atmosphere. This began with tests in the Nevada desert as part of the Manhattan project to build the Hiroshima bomb, and continues to the present day in various forms. Alongside the ongoing problem of civil nuclear waste and its contested disposal, nuclear politics will never leave us even when we want to forget it as the unintended spawn of technological hubris. The nuclear age, the material products of which are a primary Anthropocene marker, is not dead and will not be for some time.

John Hersey's groundbreaking journalism on the survivors of the first ever use of nuclear weapons – the eponymous *Hiroshima* – was serialised in the *New Yorker* and based on meticulous reconstruction of the hours and days after the US bombing of Japan. Forty years later Hersey returned to Hiroshima to capture nuclear warfare's long tail. In Hiroshima we find the same narrative trap as Chernobyl, and just as we are familiar with the drama and devastation of the weeks and months after Chernobyl, both Ukraine and Japan have been relegated to historical anomalies and not things with enduring long and entangled effects. Brown's perspective has come in for some criticism from scientists dismissing her historical ethnographic approach to nuclear legacies, but hers is not a critique of science as such, but of the tendency to look for specific events and their resolution over the long entanglements that the Anthropocene binds us into.

Nuclear disaster is far from the only Anthropocene archetype. Climate change, which has been characterised first as theory, then as risk and finally as reality, is not a straightforward story. As the environmental journalist Elizabeth Kolbert describes it, it is 'The worst story ever. It's ubiquitous, but very hard to pin down. It's being caused by everyone and everything. It's sort of everything and nothing' (Kolbert and Roberts 2017). Climate change is not a process which can simply be turned off and resolved, even if every single country were to get up tomorrow morning and suddenly have at its disposal zero carbon technologies in every area of the economy. Climate change is everywhere and there is no shortage of coverage on climate as a topic, but how it is covered can fall back on specific modes which, like nuclear disasters or environmentally driven public health disasters, mask the fact that the processes are reflexive, ongoing rather than finite, and that the usual modes of cause, effect and resolution we envisage in politics and economic management do not work for problems of this scale.

A prime example of this type of thinking is the discussion and reproduction of 2 degrees Celsius as a cut-off point for global warming to avoid catastrophic climate change. The 2 degrees target – which as of 2024 the world is still on course to miss by a considerable margin – is an agreement between nation states, but also heavily influenced by the history of climate change in the social space. Emerging in the 1990s, warnings that 2 degrees Celsius was the maximum allowable temperature change followed recommendations originating from the work of the American economist William Nordhaus in the 1970s (Jaeger and Jaeger 2011). As long as climate change remained an abstract future risk the targets were a focus for discussions of how to decarbonise the global economy rather than meaningful in themselves, as the scenarios they described were nowhere near being reached at the time.

The 2023 IPCC report made clear that not only were we as a planet on target to miss the 1.5 degrees of warming in the supposedly groundbreaking Paris Agreement, itself a politically negotiated package, we were also some way off avoiding 2 degrees of warming. Moreover, even 1.5 degrees of warming entails significant and longstanding climate impacts, many of which are already being felt. Whereas in the 1970s with the right political action it would have been realistic to flatten the curve of emissions and warming to below 2 degrees, the 2 degree target has stayed around, functioning as a rhetorical failsafe measure even as it has floated closer to becoming a reality.

Calls to action and spectacular reports of calving icebergs or of water surging through the New York subway play into a genre of seeing the Anthropocene crisis as discrete events which can be turned off as quickly as they arrive, but these are events which have been going on for some time and will continue to do so. The climate crisis will last for at least a hundred years even if we were able to slash global emissions tomorrow.

It is not overly dramatic to say that every car journey, plane taken or carbon sink disrupted today will have impacts lasting for centuries as part of this cumulative damage. Although there is no shortage of long-form elite journalism unpacking the future or discussing what the Anthropocene is and how we got here, the fact that each banal act of everyday life is implicated in these Anthropocene processes is notably absent. Ultimately this manifests as a search for meaning, and long stories in short and often banal events.

Meaning in the Anthropocene: modernity's revenants

If we see the Anthropocene as intimately linked to modernity we can tell stories about it at many levels, far from the spectacular events of Ukrainian nuclear sites or South American rainforests. Anthropocene conditions are both slow and fast burning, and increasingly these conditions result in events that break through in spectacular ways, at sites far distant from the environmental flashpoints we might usually focus on.

Scotland, as a small country in northern Europe with a lot of its land well above sea level, has generally been regarded as distant from the worst current and potential impacts of climate change. With a temperate climate and a small population relative to its landmass it also possesses generous renewable energy resources. Scotland is also an interesting case because it makes claims to being one of the cradles of European industrial and intellectual modernity as both an Enlightenment centre and as one of the early pioneers of fossil capitalism. As an early industrialiser, and driven by plentiful supplies of coal, it quickly developed networks of infrastructure linking mines to industrial urban centres and urban centres to one another. As a consequence, much of Scotland's 'modern' infrastructure is actually between 150 and 200 years old and its economy is now distinctly post-industrial when compared to its place as a centre for manufacturing and as a sizeable historic producer of greenhouse gas emissions.

On 11 August 2020 Scotland experienced freak rainfall, even by the wet and wild standards its Atlantic weather system warrants – in places 30–50 mm or more of rain fell within three hours or less, causing flash floods and landslides. This matched a pattern of increasing heavy and sudden rainfall predicted by climate models. Roughly midway between the cities of Edinburgh and Glasgow the 180-year-old canal that was one of the early conduits of the Industrial Revolution took on so much water that it breached, exploding sideways and sweeping away a large section of the main rail line between the two cities. Meanwhile, in the north-east of Scotland a morning train from Aberdeen – centre of the UK oil industry – to the south hit a landslide caused by the rainfall. It careered off a bridge and down an embankment before the diesel locomotive caught fire. The fatalities were only in single figures because the service was nearly empty on account of the ongoing COVID-19 pandemic.

The country found its two most important rail corridors impassable for months, but more interestingly the media response quickly began to identify climate change as a significant driver in both disasters. An interim report by the British rail infrastructure management company said that not only had climate change contributed to the disasters, but it would continue to do so in future, opening up the possibility that the disasters were not finite events but representative of larger long-term processes. This was carried by the

BBC and London-based newspapers such as *The Guardian*, as well as by the domestic Scottish and regional press. Climate change was presented not as a moral imperative in Parliament amid warnings from scientists and activist groups, but as an event happening in the dark hours of the early morning, striking at the heart of the country, and a new cycle of cause and effect was established in the media.

A day later the Scottish edition of *The Times* carried the headline: 'CLIMATE CHANGE WARNING AFTER SCOTTISH RAIL LANDSLIDE' (Horne 2020). The claim was repeated in other forms elsewhere, and as the story progressed there was clear attribution of the events to climate stress, which opened up space for wider conversations about the challenges of extreme weather in the future.

As ever, the question of cause and attribution is a very simple case of asking what happened and why. Superficially, there was a rainstorm, but it is possible to trace the cause of the event to the cumulative effect of processes begun when the first steam trains began hauling coal across central Scotland in the 1820s, representing the irony but also the reflexivity of the situation. All events that take place in the present are by definition Anthropocene events – any process that uses carbon or involves us taking things, using them and ejecting them out into the world constitutes and accelerates the processes of the Anthropocene, but certain Anthropocene events are imbued with meaning significant enough that they break through and assume a higher level of importance. Since the 1840s many trains have crashed or come off the rails, causing death and destruction,[1] but in this case climate breakdown became part of the semiotic of the industrial disaster. This was not only the idea that freak weather had contributed to the tragedies, but that this freak weather represented an increasing trend towards climate impacts on critical infrastructure, backed up by data and with sobering realism.

The actual trigger for explicit recognition of climate aspects in this case came from an official report from a government agency, which then gave journalists licence to run with the climate change headline. Combined with the sheer shock value of the parallel disasters and the need for explanation, climate change emerges as an agent in the story where it might otherwise have sat more obliquely. Moreover, many in the country had experienced the extreme rainstorms themselves, so the attribution of the disasters to climate change also had a strong phenomenological component. Quite simply the news was telling people what they in many ways already knew and had seen first-hand, that the storms in question were part of a series of unusual and extreme weather phenomena in the region which gave a taste of the climate disrupted future and showed the potential for infrastructural chaos.

In the near term we can expect large amounts of infrastructure worldwide to come under increasing pressure from climate effects, be that flooding, buckling or overheating due to unplanned-for temperatures, or disruptions to supply chains and labour markets far removed from our own locales. There are multiple Anthropocene news events that intrude on our lives, with the COVID-19 pandemic being a prime example of Anthropocene problems beyond climate. A virus which seemingly originated in bats, and which was spread through human interference in fragile habitats, both exposed the precarious nature of global economic systems and served as a reminder that societies are still very much bound by 'nature' and our interactions with it. Climate change, pandemics and phenomena such as antibiotic resistance remind us again that we have not escaped nature. It is possible to find cause in these effects, and linking one seemingly disparate process to another is a critical part of forming a holistic account of the complex material consequences and flows of the planet at any particular time.

Locating the Anthropocene

A flooded railway line, a house subsiding and a commodity price surge are all Anthropocene events, but we do not name them as such. Even when the terminology of the Anthropocene does not feature heavily in reporting, or beyond the pages of elite magazines and weekend supplements, it is always there and to different degrees we internalise its logic. Most of us are familiar with what it is like to live under some form of capitalism, despite capitalism itself rarely being named; we think about it most often merely as everyday life.

Just as capitalist modernity is reinforced through our media consumption, so is the Anthropocene. Local newspaper stories consisting of hastily rewritten press releases about low-cost airlines starting new routes sit alongside reports of flash flooding, and debates about the legacy of European slavery are followed by news of the withdrawal of the United States from an overseas deployment just as a hurricane hits Louisiana. This is a long way from the TV reporter in winter survival gear standing in front of a calving glacier, or the reverent and elegiac tone of an article on a threatened species in the *London Review of Books* or the culture feuilleton of *Die Welt*, but these are the places we experience the unfolding of the global present and the revenants of the past.

The Anthropocene can be useful to journalists trying to weave together this complexity of being that defines our everyday lives, but as a typology for the public it risks falling into the same semantic long grass as sustainability or globalisation, world-spanning concepts which are flexible and nebulous. As it ages as a framework for our understanding of humanity's precarious position it can also provide a means of challenging orthodoxy by clearly pointing out the entanglements of everyday life and everyday media. This is important not least when talking to lay publics. As the urban studies scholar Karen Bell (2020: 139) suggests in her work on working-class environmentalism, many working-class and non-elite people already implicitly understand and experience the negative structures of capitalist modernity and do not need to be 'won over' or to achieve the same state of mind as elite activists and commentators in order to 'do' environmentalism. The same can be said for journalism in its diversity, in which core characteristics of the Anthropocene concept can manifest and exist across a variety of types of journalism without ever being named.

In the old journalistic adage, 'show, don't tell', it is quite easy to show what is happening by being inventive with form and composition. Only small changes are needed to show grand processes, but equally by aiming low and abandoning some of the bombast of the Anthropocene narrative we can pick out its most banal and common aspects. This again is where we find John Hartley's conception of journalism as modernity's textual system and sense-making practice useful. For the Anthropocene to be meaningful as a sense-making practice it has to be operationalised within the wider everyday experience of capitalist modernity. This is not something that journalists can simply determine to do, caught as they are by the pressures of editorial demands and financial bottom lines, but the consciousness of global events, histories and futures is a composite process and not a radical new epistemology. This means that the Anthropocene is not an object to discuss but a process, and above all perhaps a lens to bring alive the myriad happenings of the global present.

4

Media Hot and Hotter

In the early days of mass internet connectivity there was a running joke about having 'completed the internet'. This was partly a wry nod to the fact that the internet had already surpassed the amount of content one person could read in a lifetime by the turn of the millennium, but also reflected a deeper trend of people living within certain self-made boundaries despite the sheer scale of what was on offer to them. In a world of endless opportunity, and in a pattern that proved prescient of the eventual platformisation of the virtual space, users would visit a relatively small number of specific sites to meet specific needs. The early internet was in many ways a facsimile of the real world, complete with websites that sought to function as little more than information boards for materially existing objects, or in some cases individual pages that existed as a space to upload and visualise content designed for physical presentation. Prior to the cascade of social media, users would construct personal websites that functioned as photograph albums, family newsletters and public diaries, often loading in pixel-lines with no sense of how important speed, volume and airbrushing would become in the next generation of online personal presence.

Many early attempts at online journalism worked on this assumption too, seeing online news as a digital visualisation of their current output. Before the advent of tablets and smartphones, the management of *The Guardian* is said to have toyed with providing its readers with printers so that they could print out the paper each day at home, while early experiments in digital distribution amounted to sending customers pdf files of the print newspaper. Even today, the digital subscriptions of many legacy titles offer a digital version of the printed newspaper for reading on tablets, as a nod to the idea of a newspaper as a distinct material object and each day as a separate chapter in the news cycle that starts and finishes as we open and close our eyes. The temporal and technological limits of news have always been evolving as both technologies of production and modes of consumption have changed, and the idea that the internet is a clear dividing line between the old world and the new is only part of a wider story of emergent technology to which it provides the glue. If you set out for a job in most arms of journalism today, you inhabit a shadow of what used to be in terms of institutional resource, and yet paradoxically you have more resources available than ever before at ever cheaper prices.

In the early 2000s university media departments educating journalists and media workers would still have banks of mixing desks, radio booths, TV studios and other specialist equipment that most journalists will now never see — more likely their workplace (if they have one) will have a soundproofed meeting room with cameras and streaming microphones. After the expensive studios and galleries came computer production suites, with digital cameras and specially purchased high-powered computers running production software. Now students can edit and master content at home using laptops and cloud-based editing suites, shooting on smartphones in higher definition than news cameras did ten years before, while people shoot feature documentaries on handheld digital SLR (single lens reflex) cameras.

In an industry with a constant appetite for saving money, these innovations more than compensate for any slight reduction in quality. During the COVID-19 pandemic many journalism schools and media organisations simply switched to using broadband connections and smartphones to produce broadcast-level news. Journalists were told that they had all the tools they needed in their pocket, and that narrative has now forged its own reality as journalism tries to pick up the pieces and refuses to revisit older, more expensive ways of working.

It is easy to see this trend as part of a straightforward story of modernistic efficiency — the technology of media has become better, smaller, faster, and also cheaper. Radio, which has long been a more mobile sibling to the expense and prestige of audiovisual (AV) content, is now as simple as a smartphone or digital recorder with an appropriate microphone to give the desired sound quality. Students of media are often surprised to find out that the limiting factor on the quality of their voice is not their phone itself but the cheap speakers fitted to most devices. Even with modest investment in equipment, for a few hundred euros you can now record a professional quality radio piece from anywhere in the world with an internet connection and somewhere to plug in a laptop. Media technology has become so easily available that — for better or worse — almost anyone can set themselves up as an AV journalist, and many have done so. This is not, though, merely a question of doing the same things more cheaply, but of opening up new avenues for reportage.

Media workers today are often expected to produce for more than one medium, to write text, script, edit audio and camera rushes, and to provide all of these to their final outlet. Automation and the use of AI tools has made this easier, but these efficiencies have merely accelerated the drive towards microlabour. There remain well-resourced news and current affairs teams with larger budgets and dedicated teams of producers and editors, but these are increasingly the preserve of an elite of broadcasters who are unrepresentative of the average media worker. Technological convergence has significantly blurred the boundaries of media work, and this shows no sign of abating.

Some more financially stable news organisations with the money to experiment have shown a curiosity for the more ambitious aspects of this as a proving ground for immature technologies, such as in 2015 when the *New York Times* sent subscribers free self-assemble cardboard goggles into which they could insert their smartphones and experience stories in three dimensions (3D), as well as generating an augmented reality 3D news stand based on a scan of a real installation from New York. The news stand appeared in people's homes as an example of the potential for high-definition 3D visualisation from the comfort of their sofa, albeit still clumsy and inconvenient as anything other than a novelty.

Hybrid formats and sensory expansion

It is easy to mock the hubris of such projects, promising a projection of reality directly into people's sitting rooms and the realisation of a dreamscape straight off the pages of 1960s sci-fi magazines. More importantly, such use of augmented reality (AR) coincided with the mainstreaming of the technologies more generally, and there are multiple examples of in-depth stories being created using AR tech in which users were invited to visit a refugee camp, take a trip to Mars or see US Olympians in 3D. The main use of AR is thus an effect previously achieved through narrative journalism and careful description, breaking down barriers of time and place to bring experiences to the breakfast table and the living room sofa. Around the same time that the *New York Times* was rolling out 3D news, Google released an AR visualiser to render CGI animals in real proportions standing on your bed or growling in your kitchen. Like many technologies of the burgeoning projected internet, its ultimate worth depended heavily on the appetite of the user to indulge the platform's idiosyncrasies. Arthur C. Clarke's maxim that any sufficiently advanced technology is indistinguishable from magic might not quite hold true for AR, clunky as it remains, but it still has road to travel.

More prosaically and practically than the pixel animals and renderings of incipient AR, there has emerged a trend towards immersive storytelling using digital storytelling tools which aim to reclaim or reinvent the art of long-form journalism. Media outlets worldwide have experimented with digital storytelling to cover gang wars in the Philippines, the cracking of the Thwaites Glacier in Antarctica, animal migration in Africa and sea level rise in Bangladesh. These stories, which ordinarily contain text, high-quality images and some form of additional video or audio, take advantage of their ability to play with form by using the page itself as a moving theatre set to convey space, time and progress. At a still less glamorous scale, we now routinely see audio, video and text presented side by side as part of the current affairs output of multiple channels. Despite the failure of industry trends such as the infamous 'pivot' to video that swept through the global news media in the mid-2010s, rumoured to have cost countless journalists their jobs thanks to a moment of technological hubris (Owen 2021), hybrid models of delivery are commonplace. This digital and hybrid journalism is now mainstream, and nondigital journalism has increasingly remained only as a form of artisan practice.

This array of tools perpetuates something which characterises the contemporary media landscape more widely, namely the phenomenon of noise. We are drowning in a world of sounds, experiences, noises and stimuli in which news consumption is a question of filtering out as much as tuning in. Marshall McLuhan's (1964) original division of media as hot or cool has been superseded by a warming media world of hot and hotter media, in which the stream of information (and disinformation) and AV content is an exponentially growing torrent that looks very much like the hockey stick graphs of emissions or energy use plotted over the past two decades. This richness, this medial heat, is still by and large something positive, and our ability to capture, edit and redistribute the world around us provides new avenues into hearing, seeing and feeling the Anthropocene in ways which can be both overwhelming and emancipatory.

The everyday tools of multimedia journalism are familiar to anyone who has ever checked the weather app on their phone or made a panning shot of a beach for a family group chat. The foundational device is a smartphone with a data connection, an audio recorder, and for some a microphone and a camera, all supported by a laptop and a suite of editing software for the more seriously inclined. We can plug-in, edit and complement these technologies through further kit, and when capturing our environment we can add to this though hydrophones, Geiger counters, particulate counters, UV (ultraviolet) meters, drones, seismographs and an array of other tools which do more than record the voices and movements of people alone. An explosion in so-called 'sensor journalism' (Caswell 2016: 1145) has meant that journalists can set digital bear traps, waiting for stories to walk into their nets of measurement, and offer up insights that would otherwise have remained invisible. Linked to data journalism in its scope for conveying scale and nonverbal narrative, sensor journalism is at its root the expansion of the sensory capabilities of the journalist and the automation of processes that would previously have required a massive labour force in terms of researchers and distributed data collection.

With sensor journalism the interest in the nonhuman aspects of the planet previously confined to nature writing or the spectacle of the nature documentary is suddenly more tangible. If you want to find out what Chernobyl is like beyond the retold tropes, you can measure the distorted scratching of the radiation levels, listen to a glacier calving to realise that ice is melting in real time, or record and play back the ambient noise levels across a city to show the cumulative impact of traffic on its residents and wildlife. Journalists have long made the news rather than merely gone and found it, so why should opening the gaze wider still be any different? Media consumers are unlikely to benefit from day-long recordings of the creaking of ice outside of the most artistically minded forms of environmental mediation, but inserting five seconds of the creaking ice into a news report moves the collapse of the world's glaciers from a scientific statement to an ongoing and tangible experience.

Wildfire media and the new bricolage

The forest fires which began sweeping North America and Australia in the early 2020s held an affective power for audiences around the world due to the widespread presence of sensory media tech for the first time, with the displaced people having access to media technologies that generated a constant stream of recording. Smartphones relayed smoke and flames as they happened, and footage of people being driven into the sea made its way onto social media almost immediately. Much of this material is raw and unfiltered; there is little attempt to craft a narrative, and indeed to do so would be a touch distasteful given the gravity of the situation. In providing people with smartphones around the world, we have inadvertently created a panopticon with a reach beyond the wildest dreams of the embryonic global rolling news broadcasters of the 1980s and 1990s, spilling onto the internet in real time. There is now a camera crew on site to capture anything within spitting distance, and these sights and sounds expand quickly to audiences eager for the spectacular and the immediate.

Although wildfires are by no means the only phenomenon captured by these networks of handheld media and the associated social spread, the processes involved have been analysed in terms of smart and digitally mediated landscapes. The media sociologist Jennifer Gabrys, for example, has written extensively about acute forest fires and their mediation, centring on the use of smartphones and other forms of everyday sensory media linked to a broader understanding of global climate change and extreme weather (Gabrys 2022; Gabrys et al 2022). The same phenomenon was clearly visible in 2022 during the catastrophic floods in Pakistan when footage forced its way into the media agenda of the rest of the world due to the sheer prevalence of shocking content filmed by Pakistani citizens. With climate change worsening, these live and almost-live encounters with the catastrophic can only be expected to increase.

Such videos reveal networks which, in Thomas Pringle's words, are 'situated alongside regionally-based collectives of individuals soon to be thrown into uncertain political configurations in a changing and warming world' (Pringle 2020: 38). On the continuum of professionalism from full-time salaried journalists to bystanders via community or citizen reporters, the immediacy granted by such handheld technologies exposes the complex infrastructures and networks already in place, but also calls into question the notion of journalism as a professional practice in which meaning is subsequently tagged on in the participatory space of social media.

This is a process replicated across other global stories as well. The complete collapse of US military strength during its withdrawal from Afghanistan in 2021 was seen equally through the eyes of professional international broadcasters and locals. Videos of children being passed over airport security fences, or the Taliban going door to door with guns looking for political opponents, were better than any news producer obsessed with the right shot could have hoped for. During the invasion of Ukraine by Russia, mobile phone footage, selfies, drones and CCTV all fed into a montage of the present that anyone could dip in and out of on social media. The carefully choreographed set pieces that formed the basis of *Our World* in the late 1960s have given way to a chaotic naturalism in which we can sample organic events as they happen and assemble a bricolage of experience.

Wildfire tracking is an obvious and tangible use of social media and mobile technology, but the other calling cards of the Anthropocene can be documented just as well – not least sound.

Sound and place together create new engagements with space. Sound helps us to see and feel weather in media, particularly what the anthropologist Tim Ingold (2010) terms the 'weather world' (the spaces through which we move and which interact with us through sound, sensation and, perhaps most of all, location). The richness of sound can capture landscapes and place better than any other medium. Something as simple as a radio recorder with X/Y 180 degree mics aimed at a street will capture everything from the murmur of bus engines to the rustle of trees, passing aircraft and children playing on the school tarmac, bypassing the human tendency to focus only on one or two sounds at a time. The sound of a rainstorm, isolated and played back, creates a *mise en scène* for the report itself. The same can be said for metal roofs clicking in heat or high winds flashing past.

Even less extreme Anthropocene events can be rendered tangible with sound. The journalism of sound allows the environment in all of its overwhelming complexity and messiness to be felt, yet journalism seems reluctant to engage with this sonic power. Geographer Anja Kanngieser (2015) has written about how sound allows new interrogation of inequalities, imperceptibility, translations, commons and futures, describing how soundscapes reflect the changing distribution of power and governance through their aural and cognitive cues. We even know that changes to soundscapes can have highly detrimental impacts on the environment and can change habitats and ecosystems, as shown for example by what sound researchers have termed the soundscape of the 'Anthropocene ocean' (Duarte et al 2021). Sound is not just representative but constitutive of the Anthropocene, and using tools of sound in journalism allows us to more carefully interrogate the sensory components of the human and natural world alike.

These sights and sounds are not just the nature documentary staple of capturing what Jamie Lorimer (2015) has described as electric animals – charismatic animal actors carefully produced and reproduced in lounges far away from their homes to fit cosy narratives. On the contrary, the attraction of sensory technology to journalism is its ability to go off script, and to offer us the alien and unexpected as much as the cosily anthropomorphic. Ice, the low rumble of motorway traffic and – as Rachel Carson so presciently described – even the sound of silence is where we can catch glimpses of the systems that surround us.

Technology, mobility and modernity

Technology is much more than a narrative tool for media – it defines and regulates journalistic work and dictates forms of labour. The modern newsroom with its banks of open plan desks and journalists busily producing content on screens is a halfway house between reality and audiences, in which journalists themselves are subject to levels of mediation before they engage in the production of mediated reality themselves. In the first and second generation of digital news, journalists were increasingly condemned to 'screenwork' (Boyer 2013) in which their world was shrunk to the scale of an office, and their eyes and ears were merely email inboxes, news feeds and telephone lines.

Old habits of going out to physically call on interviewees, or even attending press conferences are replaced by live links and video chats. Even the infamous death knock of British tabloid journalism – in which rookie reporters would be delegated the task of finding the relatives of people tragically killed, by visiting their homes – has been replaced by social media harvesting of Instagram pictures and screenshotted dedications to showcase the tragedy.

Conversely, the technology that initially led to the rise of screenwork as a static practice is now facilitating new forms of mobility in novel ways. Contemporary journalism and its technological faculties can be seen as part of a new 'politics of mobility' (see Örnebring and Schmitz Weiss 2021) in which journalism participates in broader patterns of mobility and is itself reconstituted by mobility. This mirrors what the sociologists of mobility Mimi Sheller and John Urry (2006) termed the 'new mobilities paradigm', in which transport and communication converge, and mobility is less a temporary state than a permanent one, but where this mobility is also bound up with sedentary habits as we move in cars and planes or 'move' while sitting still. Mobility is also a value which journalists are expected to embody, together with more figurative ideas of flexibility around time and labour (Hinde 2021).

Although it does not show up directly on the Anthropocene charts outlined by Will Steffen (which measure material and socio-economic processes), mobility is a prime characteristic of modernity after the Great Acceleration (Steffen et al 2015). The Great Acceleration, with the growth in international tourism and business mobility, has made itself felt in terms of the mobility of journalistic practice as the cost of air travel has plummeted and staff have been increasingly drawn into models of flexible and mobile labour. This has all facilitated work in an international and often transnational context, limited only by the visa regimes of states journalists may travel to. This has also been mirrored somewhat in the transnational reach of publications – in Britain, for example, multiple media outlets have set up North American operations, which take advantage not only of a shared language, but of the development of a transatlantic media industry tying together the hubs of London and New York. These relatively minor undertakings – often consisting of just a few employees – rely on mobility of labour but also individual mobility and the technological webs in which journalists are enmeshed. This is all underpinned by transnational webs of staff and freelancers paid and funded through fluid mobility of capital. Whereas *Our World* required hundreds of people across the planet to carefully coordinate its studio-bound contributors, it would now be possible to recreate it in the space of an afternoon using tools most people have in their pockets.

In this sense, global journalists live the Anthropocene, moving across the face of the planet while entangled in a web of interaction with others, using the toolkit available to them. The economisation of journalism and the convergence of defined roles into one person also has the effect of giving that person unparalleled control over what they do and new forms of flexibility. Just as desktop journalism heralded the rise of screenwork, mobile and laptop journalism has created the potential for taking journalism production beyond screenwork and back out into the world.

The return of fieldwork is an unexpected opportunity, even if that mobility still comes at the price of operating in the insecure systems of transnational journalistic business. In crisis situations in particular, journalists can report on the ground with an immediacy that was previously the preserve only of either long-term embedded correspondents or well-resourced international news teams. Particularly when it comes to questions of climate and environmental crisis, the farthest reaches of wilderness are no longer a mystery thanks to the ability of journalists to piggyback on the international travel market and the explosion in portable broadcast and reporting technology.

Anthropocene mobilities also mean coming to terms with human flows and forced movement as climate change begins to bite seriously. Building on Sheller and Urry's (2006) theory, the geographer Andrew Baldwin has argued that the mobility paradigm is foremost a style of thought, one that in an Anthropocene context takes on new meaning with the realisation that our mobility as individuals and as a species is intimately bound up with the Anthropocene concept (Baldwin et al 2019).

What this ultimately means for our engagement with the environment is an ability to develop truly cosmopolitan modes of reportage and coverage, unburdened by the excuse that cultural and geographical blindness are the result of financial and technological constraints. At the same time, we might be mindful of the fact that modernity has internalised mobility as a hallmark of progress and modernisation, and that this mobility is having severe impacts on the societies we seek to report on. The Anthropocene is always moving, and so are we.

Media technology as a collage of modernity

Through new mobilities and new technologies, engaging with the tools of Anthropocene journalism offers tangible advantages and possibilities. Experiments with form allow journalists to try and capture the enormity and diversity of Anthropocene processes, though these encounters can remain trapped in structures of media devoted to both conventional storytelling and the pursuit of reductive approaches to information transmission. The spaces for this experimentation are at present usually limited to the fringes: public service radio and television outside of mainstream consumption, or as pilot projects in large organisations who test and then ultimately decline to pursue these newer models for fear of financial consequences or due to a judgement that technology is unsuitable or unready for wider consumption. Yet the fact remains that while there can be resistance to new and more expansive forms of journalism through technology, we live in a world hotter and hotter every year, with more noise to hear and more change to capture.

Audiences are also able to build affective experiences from multiple technologies and sources without the direction of a single outlet, and will indeed do so whether or not these textures and sensations are channelled to them from legacy media or otherwise. This bricolage of experience, to use the description suggested by Mark Deuze (2006) for our consumption of digital culture, is not owned by any particular part of the media, but all parts of the media are subject to it. Affective interventions using this technology are reliant on editorial and curatorial judgement as much as the ownership of particularly directed and impressive gadgets, but listening to and documenting modernity's every tick and flex gives us unparalleled awareness of its flows as we jump between textures and sensations. This is less a question of journalism as modernity's textual system than of journalism as modernity's sensory filter. This is a world in which screenwork combines with fieldwork and where sensations can come and go in an instant, the challenge being to use technology to assemble them into a collage capable of retaining meaning.

Technology in Anthropocene journalism, then, comes down to these questions of constant movement, of redefining finite processes as constantly flowing, moving and in change. Ultimately, new news technologies allow journalists to experiment with the breadth of form available to them, limited only by the willingness of audiences to embrace this richness and of employees to pay for it. In the noise of media, embracing it and structuring it gives us the opportunity to make sense of Anthropocene processes. Now with a hundred broadcast quality smartphones aimed at a burning hillside, we really are living through the age of media hot and hotter, fluid and mobile.

By acknowledging the movement and change inherent to it and technology's ability to capture and sense these transformations, the Anthropocene is tangible as more than an abstract idea, and takes on concrete form as an ongoing process all around us.

5

Plotting the Anthropocene: Digital Modernity and Charismatic Data

The Anthropocene owes its existence as a scientific – if not intellectual – concept to a mass system of data collection and analysis. From weather modelling to the detection of radioactive particles in the atmosphere and the crucial parts-per-million (ppm) quotient of carbon that rises each year, the abstraction of world-spanning systems is rendered comprehensible through an ability to map and quantify highly complex and interdependent processes across time and space. Many of the ideas behind the *feeling* of the Anthropocene may be old, but its suggestion by Crutzen and Stoemer (2000) and its (ultimately failed) definition by the International Commission on Stratigraphy (see Zalasiewicz et al 2010) are the product of meticulous collection and evaluation of data in the past two decades.

This capacity for measurement and mass processing of information has implications for how we understand the complexity of natural systems and the impacts and behaviour of human systems, and ultimately for our ability to understand and potentially act on our own future. All of this information can be mediated through journalism and its adjacent practices. Despite the contemporary enthusiasm for data journalism, data is nothing new to the field – journalists have engaged with quantitative information for hundreds of years, including in the shipping forecasts and weather reports that functioned as early forms of environmental reporting. Similarly, provision of information on farming yields, sport fixtures and the stock market have all played a crucial role in journalism's evolution, and in each case newspapers and media outlets have sought to present, analyse and contextualise data sets which tell stories about the wider world. At the same time, media operators have long relied on data to drive the work of journalism itself, selling viewers and readers to advertisers and engaging in constant mining of the public for information that drives their business models and audience engagement.

Today, data has assumed its own semiotic as a social object in and of itself, and increasingly as a manifestation of power. With the advent of what has been described as 'digital modernity' (O'Hara 2020), data has become a lens through which to view modernity itself, with far-reaching sociological implications. This means that dealing with data is not only a question of seeing what is happening in the world, but of reflexively understanding it as a significant determinant of contemporary modernity's (and particularly contemporary capitalism's) form.

Statistical empiricism and big data are also integral to the Anthropocene's ability to churn and transform the planet. It speaks to the hubris of digitisation that claims are now made about the transcendental potential of mass data gathering and deployment, but the relationship between data and the state has been integral to the course of the Anthropocene from its roots in Victorian empire and reliance on armies of statisticians, cartographers, mathematicians and economic planners.[1] Totalitarian societies carried out programmes of mass surveillance long before CCTV, facial recognition scanners and computer databases were invented, and the global financial system existed before stock markets became dominated by the immediate and omniscient technology of the internet, instant financialisation and algorithm-driven trading.

Mass data collection and use goes hand in hand with modernity as a world system, both as an emancipatory tool and as a reservoir of power. In our current moment of environmental crisis, data is also a means by which some argue we might save the planet and ourselves through an optimisation of social life and resource use, a vision which sees data as part of an Anthropocene 'infrastructure' (Edwards 2017) that enables solutions to be found to modernity's externalities. This narrative pushes the dream of an optimised society in which the biosphere and technosphere become part of one single and balanced assemblage with humanity at its heart. There are good reasons to doubt this view, due to both the tendency of anticipated technological transcendence to manifest in more banal and ultimately unsatisfying everyday forms, and also the complex politics which surrounds data creation, its control and its use.

The digital sociologist Karen Gregory (Daniels and Gregory 2016: 5) notes how 'digital technologies, even in their most banal or quotidian form, speak to issues of power and the relationship between our individual biographies and the larger tides of history'. In the landscape of data-driven and data-mining journalism and wider society, how we see and feel the world is determined heavily by these entanglements. One of the things that characterises the Anthropocene is the expansion of increasingly complex and far-reaching modes of information collection on a material level, but also the attachment of value to this data as it enters politics and commerce and offers us seemingly utopian agency.

This belief in data as saviour, of optimising our way out of the hole we have dug for ourselves, relies on linear faith in empirical rationalism and the conviction that we need only look at the numbers and facts of the situation to find solutions to its problems. This is a world in which to 'have the data' is a guarantee of correct decision making, divorced from the context in which such decisions are made. As Alexander Zevin (2021) has argued, the projection of liberal objectivity and information-driven economism in financial analysis – of 'looking at the numbers' as a counterpoint to supposedly ideological interventions – has been a longstanding tactic of financial journalism, embodied in *The Economist*'s anonymous bylines and the impersonal objectivity it implies. The world of data thus plays on old and established ideas about rational sensibilities, whereby the ability to take stock and accurately predict the future grants both power and financial advantage. Claims to information and objective views of the world not only diminish the power of (potentially correct) fringe ideas or alarmism, they also cement power in ways which are subtle and enduring, and which in many cases help to legitimise existing conditions. This is the web of power and information that Anthropocene journalism must navigate, using data as both tool and context.

Most large news organisations now employ data journalists, and increasingly freelance and smaller scale operations are required to use data journalism in their everyday work. What began as so-called computer-assisted reporting (CAR) has evolved into a more dynamic practice of obtaining and using large data sets as an integral part of storybuilding, in line with a generational shift in the internalisation of media and data into journalists' own lives. CAR has existed in various forms since the middle of the 20th century, with the increasingly available processing power of computers used to aggregate everything from election data to weather reports quickly and accurately. Since the 1990s, however, CAR has diversified and developed into strains of programmer journalism, open source journalism, computational journalism and data journalism, among others (Coddington 2015: 332). This trend runs counter to a professional journalistic ethic in which a primary concern has been to produce copy or specifically textual or visual components, an ethic in which number crunching and data literacy have been derided compared to the traditional domains of journalistic production and the craft of written copy.

Good data journalism and literacy about the sociology of the data society can help journalists to find stories that are often hiding in plain sight, or to make sense of information which contains important and newsworthy points but they may not be immediately obvious. At the heart of this is an ongoing and uncertain relationship between data and large volumes of representational information on the one hand, and the material reality of the world at large on the other. Stock market crashes are fundamentally crises of spreadsheets which are representative of a material process happening somewhere, or in the case of futures markets things still to happen. Ironically of course such immaterial representations have real world consequences as the fruits of people's labour vanish overnight or homes are repossessed and values plummet or skyrocket. Data may be purely representational and data journalism interested fundamentally in the business of representation, but its distribution and presentation have real world impacts on action and inaction. Nowhere is this more apparent or more pressing than in climate science, where data is constantly generated, deployed and debated as different articulations of the future are calculated.

Reading the infinite library

Climate and ecological science has an exponential thirst for information to improve its models and to add nuance and depth to the basic grand processes of climate change. The 2023 full Intergovernmental Panel on Climate Change (IPCC) report utilised over 10,000 scientific publications in grounding its conclusions (IPCC 2023) – one of the largest literature review and data analysis exercises ever undertaken. The Sixth Assessment Report is the cumulative result of decades of research and constantly improving data collection and processing, and also the single most useful guide to the future of the planet. Without this exercise in mass aggregation and the individual studies that contribute to it, all of the things happening around us would be overwhelmingly incomprehensible and their politics intangible. The IPCC reports are clear examples of well-handled and meticulously presented data, but as data is constantly created, disregarded, manipulated, destroyed and stored until it is deemed useful, we still have to be wary about what is being said and by whom. When we talk about having the data we have to be mindful of *which* data, and just as anyone, including climate sceptics, cranks and wreckers, can draw a graph, understanding how anyone can generate data is critical to understanding its usefulness. Like Borges' Library of Babel, the 'datasphere' is so large as to be almost useless without knowing where to look, and moreover what to look for, even if there is truth to be found in the cloud of information.

The composite narratives of the library are not just an idiom – ultimately the huge wealth of informational material in its diversity comes full circle to form recognisable trends. When we represent climate change we have to try and document this information cloud and its complexity by finding commonalities, and to show how it reaches both backwards and forwards in time as well as spanning the globe right now. As you read this book, gases released over a century ago are having a cumulative impact on what the weather will be doing 50 years from today. We know this because we have good reliable data on carbon levels in the atmosphere going back centuries, and can plot them against temperature fluctuation to model the future too, backed up by increasingly robust monitoring that reinforces existing climate models. We can add depth and nuance, we can tweak and adjust, but the core story remains the same.

The macro-story of this data is the tangible and growing warming from the beginning of the Industrial Revolution to the present day. In 2019 freely available climate data from a project run by the National Centre for Atmospheric Science in the UK was used to generate a series of visualisations, devoid of numbers, showing average temperatures at various locations around the globe. Profiled under a public campaign called 'Show Your Stripes', participants were encouraged to locate the data for their own home and create their own bespoke map of climate change over the past century and a half. Each of the graphics generated from the university's visualisation tool looked slightly different, but all showed the same broad trend, accelerating in the period from the 1960s through to the second decade of the 2000s (see Figure 5.1).

The visual language of these charts is hard to misinterpret. As the bars slowly transform from cool icy blues in the 1850s to hot reds in 2020 (greyscale version in print for reference, with the full colour version online), the trajectory of the Anthropocene is undeniable wherever you are. Somebody with no knowledge of climate science or data can unambiguously understand what they mean, even in the absence of any scale on the visualisations themselves, and the semiotic is so universal that even preliterate children can see it. The stripes are not so much graphs as custom works of art. A single picture file on a computer has the same affective power as several pages of explanation, but by participating in their production and seeing how numbers become images, visitors to the site also become active data handlers.

The #showyourstripes campaign became a worldwide attempt to enmesh the visual language of long-term climate data into everyday communication

Figure 5.1: Temperature change in Glasgow 1850s to 2020 (in colour in online version)

Source: Rendered using showyourstripes.info, based on Berkeley Earth and ERA5-Land data sets. CC BY 4.0 Creative Commons Licence.

and to bring climate into personal stories. Users were encouraged to integrate the stripes with their social media as a means of starting conversations, but also to submit stories based around the data to climate journalism platform Climate Central. The stripes ended up on the shirt sleeves of the English football team Reading FC and on the cover of the activist Greta Thunberg's bestselling *Climate Book*, but also on blankets and cushions, tattoos, window stickers and pencil cases, and as figures in academic writing.

In these banal encounters with data we see how climate data becomes what science and technology scholars call boundary objects (Star and Griesemer 1989), where scientific data passes on to new groups and constituencies and gains a different utility. Few of us have the skills to intimately understand unprocessed climate data, but 'climate data' becomes an object we can relate to and use based on broader understandings of both its core processes and its outcomes.

Alongside its work on the Show Your Stripes campaign, the website that first made the visualisation tool public also pioneered an open access sea level rise forecasting tool, based on freely available climate data, which were then picked up by media organisations around the world to show that various localities were at severe risk from coastal flooding or permanent inundation. The affective power of these two relatively straightforward pieces of visualisation is obvious. Though Anthropocene effects can be complex and hard to grasp, within them there are very simple pathways to showing both global processes and local impacts by centring data not as an abstract scientific finding but as a narrative vector.

Touching hyperobjects

Underpinning the global journalism of *Our World* and Marshall McLuhan's theorisation of global media networks was the idea that information systems ultimately mimicked the physical globe and its natural system. Yet even with the increasing power of distributed computing and supercomputing to run models, the prospect of a machine model of the world is still more sci-fi trope than achievable reality.

Nevertheless, we are encouraged more than ever before to see the environment as a multilayered data set and a complex system. Despite the shortcomings of data generation as a teleological quest, data journalism and visualisation provide a means by which we can begin to grasp what the philosopher Timothy Morton labels hyperobjects – phenomena too vast or complex to understand through conventional narrative means. As Morton (2013) explains, hyperobjects are invisible but ever-present, so vast that we can remain unaware of them and the serious and far-reaching effects they have on us. In a similar vein to Bruno Latour, Morton asserts that 'nature' is a form of distancing from the deep entanglement we all have, all of the time, with these huge objects. Unpicking the slow attrition of multi-decade pollution, Morton imagines a cancer diagnosis that comes from toxins in the biosphere:

> Statistics tell me, obliquely, never able to point to a causal link, that my cancer may have come from an endocrine disruptor. Hyperobjects seem to occupy a Humean causal system in which association, causation and probability are the only things we have to focus on, for now. (Morton 2013: 39)

The world of hyperobjects, of the globe-spanning material things of modernity, is visible only through collecting the multiple microcontacts between these Anthropocene processes and ourselves. They may be imperfect, but assembling statistics allows us to see and understand things at scales that also reach into our own lives.

As the environmental historian Nancy Langston describes in her work on endocrine disruptors and the toxins in many of our bodies (Langston 2011: viii), the story of a single woman living with cancer in the Great Lakes between Canada and the United States provides a personal testimony of the horror wrought by the legacy of toxic industrial by-products from the postwar period. This story is complemented by an array of hard data which reveal that there were multiple women with the same experiences in and around the Great Lakes, suffering from the same health effects as chemicals were passed from factory to water, then to fish, and ultimately into breastfeeding mothers and their children. Like climate change, environmental contamination is often a hyperobject which can only be seen at scale. It also provides a chance for us to bridge the gap between the grand narratives of history, structure and space on the one hand, and our own lives on the other.

In the case of the Great Lakes and the toxins in people's bodies, this means data on reproductive problems from the assembled hospitals of the region, but also for farm animals, and reliable and consistent reporting of levels of contamination in the local hydrological system going back over half a century. This is contextualised through the meta-analysis of multiple studies (simplified through the establishment of coordinated worldwide digital libraries of medical knowledge in the past decade) which shows that the same process is both scientifically likely and possibly also happening elsewhere. Suddenly one woman's story becomes a node in a huge worldwide process that impacts on us all, giving a human face to a very large and complex problem. What is happening in one place is likely to be happening elsewhere too, but what happens in one place can also have knock-on impacts at a distance far removed from its immediate location. Climate change is perhaps the ultimate proof of this because it works not just across great distance but also transcends time over centuries.

Rationalism, realism and contestation

Central to the enthusiasm for data as a guiding light in the objective ethic of journalism and its claims to truth is the belief that building the right kind of future is dependent on ignoring the siren calls of ideology and focusing only on the hard facts of the situation. In the new wave of science wars that have arisen out of the global recognition of climate change, contestation of the facts of climate change has increasingly come to be dominated by accusations of poor method and ideological bias, with large resources devoted to challenging mainstream climate science in the name of intellectual rigour.

In 2009 hackers were able to access internal emails from the University of East Anglia's (UEA) Climatic Research Unit in the UK, one of the leading centres for research on anthropogenic climate change and a key contributor to both British and international climate policy. The hackers made public a series of email exchanges which appeared to show climate scientists manipulating data for ideological ends, combining different data sets to give the impression that climate change was worse than the reality.

What was not visible in the emails was the knowledge of the researchers themselves and the experience of statistical handling that informed their decisions. One particular point of contention was that the means of measuring temperature rises using data from tree samples had been combined with actual measurements of global temperatures carried out since the 1960s, a method which sceptics seized upon as evidence of a conspiracy (Skrydstrup 2013) and which sparked media interest in 'climategate' from reporters acting in both good and bad faith.

As one of the most high-profile public contestations of established climate science to date, the UEA email leak is instructive because it showed the public the human aspects of climate science, experts discussing and agreeing on findings based on their collective specialisms. Such back-end communication about how things should be presented and published is de rigueur in the editorial meetings and processes of journalism, yet journalists are always quick to refute any suggestion that their output has been editorialised or selective, or that its truth is a negotiation of reality.

The epistemological integrity of science is a field where science and technology study scholars and conspiracy theorists at first glance find common ground,[2] but the realisation that climate knowledge is coproduced with human actors belongs to the same superficial critique of social science academics who are accused of bringing politics into the classroom. Science's necessary pretence to complete objectivity is the target for people with a variety of agendas, stretching from conspiracy theorists watching cued-up YouTube videos and chatting on message boards at one end, to the powerful lobbies of the fossil fuel industry at the other.

Sceptics and deniers think – often with great sincerity – that they have stumbled on a political conspiracy, but data handling is an interpretive skill which requires some fluency and which is easy to get broadly wrong. Like the leaks at UEA, people cast doubt on or find holes in substantial and well-grounded information, and unlike the science wars of the 1990s in which empiricism found itself under attack from sociologists and antiscience fundamentalists on either flank, we now see information on controversial topics attacked for its supposed lack of empirical rigour. Using data in journalism means having an awareness of its fallibility not just as an empirical resource, but as a rhetorical object which can be accepted or rejected by publics and critics.

One of the lessons of the UEA climate case is that although there are many bad faith actors operating in the climate space, the idea that climate sceptics are irrational actors, pitched against the prophets of rational enlightenment armed with the data, is also problematic. A public arena teeming with pseudoscientific voices and ideas presented as magic bullets, feeding an appetite for speed and novelty, has created an intellectual marketplace in which high-impact statements can easily masquerade as substantial research. Policy forums, TED Talks and the annexation of intellectual space by content-hungry publishers keen to print and produce creates a grey zone which rewards bombastic claims and simple narratives while reducing social analysis to a filmable and editable pep talk in the worst traditions of Silicon Valley.

We see this in the presentation of the Great Acceleration as an idea itself – what the philosopher Clive Hamilton describes as the theodicy of the 'good Anthropocene', in which the global crisis becomes a means of human emancipation towards a more perfect and advanced society (Hamilton 2016). The war for empirical superiority and the disdain for emotion and the political is typified by the guiding ethic of concepts such as 'factfulness' (Rosling et al 2018) and the popular economics books on the reading lists of thought leaders and the financially mobile, from Davos to climate summit negotiating halls and think tank roundtables. The power of good news stories which claim to bring us to our senses and see the world through rational eyes is obvious in a time of crisis and uncertainty, yet such claims to calm objectivity are undone by the urgency of the current global crisis.

The guardians of rational debate regard themselves as an antidote to fake news, and to the perceived irrationalism of reactionary politics, but they also critique the alarmism of the climate movement and those critical of the structure of international development and organised capital. In this view of the world, the teleological mission of statistics is to present the world as a spreadsheet without looking at its interdependencies. Like *The Economist*'s impersonal persona or the calls for restrained analysis of the international policy world, statistics can be presented as divorced from important questions of power, action and context.

Christopher Anderson has mapped how data is used in journalism as both a reporting tool and a communication mode, describing how data is often used to position journalism as the heir to the Enlightenment ideal of truth and fact in spite of the contested nature of statistics themselves (Anderson 2018). This is a world in which the production of knowledge from data is deemed to have higher worth than qualitative journalism, and where journalists often do not stop to contemplate the link between the two.

In the quest for acceptable deradicalised solutions to the Anthropocene crisis, data is also presented as a driver of efficiencies divorced from politics, inviting us to look to technology and innovation over political reform. Concepts such as 'smart cities' are pitched as a necessary organisational response to the problems caused by modernity and the 'disorder' of capitalism. In the smart city, for example, we can rest assured that technology is making our lives easier and more streamlined by producing a new type of public goods through digital efficiency. The narrative of the smart city is one where you can have more efficient council services, experience less pollution (through the emancipation of self-driving electric cars and on demand transit), and find out which is the fastest and greenest route to your destination. Add to that more banal activities such as being able to wire your home heating system to the weather app on your phone and have your groceries waiting for you in a lock box when you get back from your timed afternoon run, and data and its providers become an emancipatory tool to optimise your life, calmly coordinating in a way we could not possibly carry out at scale.

What emerges is not, then, a neat division between those who accept the facts of the Great Acceleration and those who deny it, or between reason and populism. We are engaged in a worldwide epistemological conflict, but the situation is more complicated than a dichotomy between the holders of rational truth and what is popularly represented as a tide of misinformation, disinformation and 'fake news'. The attraction of a sensible, rationally driven global political settlement in which populism is pushed back from the territory of reasonable sensible politics is clear, but restrained governance is not what is needed in times of extreme planetary change; nor can we find solace in the comforting predictions of costs and investments that populate discussions of the energy transition. Climate data tell us that we should be alarmed, but the future is mapped using the self-same grammar of numbers, costings and trade-offs that demands restrained realism from governments and anyone wishing to be taken seriously.

The cruel optimism of empirical certainty

This kind of thinking is a prime example of what the late cultural theorist Lauren Berlant classified as 'cruel optimism' (Berlant 2011): the compulsion to attach dreams to rigid temporal expectations, and to return again and again to the ideas of future betterment intrinsic to high modernity and its belief in systematic progress. Berlant sees this optimism as cruel because it allows us to invest emotionally and financially in futures which are unrealisable, and furthermore that the metrics we use to measure the success of such futures are deceptive. Individually and collectively we can fantasise about future prosperity, about reciprocal recognition, about meritocracy and about the right outcome being reached through careful quantification of the future in personal loans or investment in ourselves, and of self-realisation through improvement and efficiency.

The cruel optimism of statistics and data visualisation is that we can choose to invest in information while ignoring its fallibility, and indeed construct good news stories around it which allow us to continue pursuing detrimental structures in the present. Just as, in media, owners construct narratives of reconstruction based on notional user bases, eyeballs and demographic penetration, we can say the same for boasts about the number of electric cars on the ever expanding road network or the recycling rates for airline meal packaging on the road to net zero – small trends that obscure the larger systematic failings and inertia of capital.

Indeed, when we are tasked with taking stock of the future we have to do so in a way that factors in environmental and social externalities and risks, and the reality that the Anthropocene did not come into being as a result of a nebulous and uniform 'humanity' going through an unavoidable phase, but through the use of fossil fuels to collect wealth clustered in particular corners of the globe long after the impacts of doing so came to public attention (Keys et al 2019). The trajectory of the Anthropocene illustrates the bind between wealth and resource extraction, progress and regress in its shadow, and of course the Faustian pact made between accelerationists and media innovators gambling on transcendental breakthrough.

The need to internalise these Anthropocene risks and model them against nominal measures of social progress underpins much of the work of the IPCC, both directly and indirectly. This has the strange effect – although the causes of climate change are intimately understood – of demanding that we collect even more data to provide a picture of the available trade-offs between these two sets of statistics about the balance of risks.

As journalists experiment with visualisations and statistical contextualisation, from proprietary packages to open source toolboxes, they actively engage in the curation of numbers, and are thus faced with a series of choices about what they present, how and why. Instead of obliquely signposting people in the direction of 'the facts', handling such information is about understanding the larger stories at play and how data is consumed as those stories are co-created in the minds of publics.

Finding modernity's stories in information

As sea levels rise, we feel the flows of modernity through our smartphone screens and our memory alike, constantly building and rebuilding stories about the present and the future as we are fed structure and sensation. High water estimates become more meaningful when mapped onto residential neighbourhoods; rainfall predictions take on material form when rendered in relation to existing experiences of rain or the arrival of a higher than expected insurance quote. Average global temperature rises of a few degrees sound inconsequential until the data is mapped for variation onto specific cities and regions. Data alone is not enough to portray the planetary system humanity has created in all its scale and horror. Buying into the Anthropocene as an idea, and modernity as a process, is about much more than accepting its scientific markers and factoring them into a managerial assessment of global risk.

Anthropocene measurements and data sets are not just made up of climate and environmental data. Because modernity as an object is ubiquitous, the information it presents also stretches well beyond the confines of scientific research and into the informational systems of global capitalism. Climate data may tell us about climate change, but there are numerous other types of information that help us to unpick the contemporary moment that causes it.

The emergence of open source data journalism and file sharing in the early 2000s attracted a great deal of attention from journalism researchers, and allowed journalists to recast themselves as public servants for the digital age alongside an enigmatic and ambiguous army of hackers. WikiLeaks and its fellow-travellers promised to upend the information monopolies of business and government in the public interest, largely unconcerned about the legal implications. In 2017 the Munich-based *Süddeutsche Zeitung* newspaper, together with the International Consortium of Investigative Journalists, sparked the Panama Papers scandal, revealing a worldwide network of financial offshoring and wealth accumulation in which multiple powerful companies and individuals were implicated. Once the core documents were released, experts and amateur data journalists alike sifted through them and started to make connections. One open data set provided the structuring paragraphs of a story about how global finance worked and where the rich had hidden money desperately needed by the social welfare systems and transition funds of countries the world over. Behind the Panama Papers was a clear picture of how global capital operated and what this meant for the rest of us, showing how huge wealth could be amassed by a few and made to vanish in the global financial system. By extension, this was a story about capital, modernity and Anthropocene processes. Emails between fossil fuel lobbyists, details of the contents of global tax havens, and the network maps of social media are equally valid forms of Anthropocene data as much as climate forecasts and nitrogen accumulation.

Open source hacker culture was built on the premise that only by operating in a decentralised legal grey zone could important information be brought to the public, yet much of the information that might help us tell important stories about our trajectory is hiding in plain sight; the challenge is merely to make connections between its disparate parts. We all generate data constantly, and this data is processed and transformed into statistics by two major groups – governmental and state actors looking to manage risks and strategy, and commercial organisations for whom it provides an economic advantage in the form of both market intelligence and as a financial asset in itself.

A burgeoning ethic of open governance has led to increasingly large amounts of government data being made available to anyone with a smartphone and the ability to read xls or pdf files. What we as members of the public and journalists rarely have access to is the proprietary information held by private companies about us and about the world in general.[3] Particularly useful is data that shows change over time in ways collective memory or individual observation can fail to capture.

This could be that the amount of green space available in a city has diminished substantially over time, or that governments have consistently spent more on roads than public transport, but nobody has chosen to illustrate it so far because these changes can happen in a piecemeal and disparate way despite their collective effects. Data which is covered, obscured or merely forgotten can make its way into the hands of journalists and into the public realm, contradicting public statements or policy approaches and giving journalists and laypeople power and opportunities to counter governmental and commercial interests in particular. Headlines appear asserting that 'new data' shows government failing on climate commitments or that emissions may not fall as quickly as a press officer has led people to believe at a policy launch. Such data hooks keep the climate story rolling when there are no flash floods or heatwaves to be had, and the constant feed of data-driven stories, as well as the science of climate change itself, showcases the constant turn and churn of the global present.

As the digital humanities scholar Bethany Nowviskie points out, data as narrative is a way of capturing the '*longue duree*' (Nowviskie 2015) expanses of time in which we are entangled, but also the spatial interaction of the here and now. Journalism does not habitually delve into the deep past in its search for causation, nor does it look into the far future in terms of its anticipation of consequences, but we are constantly plotting the course of megasystems and historical processes even as we sit at our desks.

Anyone can turn on their phone and see the location of any civilian flight or ship on earth, telling stories about the distribution of global wealth and power as they cluster over the conurbations of Europe and North America. When COVID-19 caused a sudden plunge in global oil demand in 2020, redundant oil tankers registered as red dots around the coasts of major producers and consumers, while the absence of air travel registered in the lack of flight numbers moving pixel by pixel over the Atlantic and Pacific oceans.

If we imagine modernity itself as the hyperobject par excellence, everywhere but often intangible as anything other than a peripheral sensation or location, finding points of contact depends on this ability to visualise networks, patterns and processes with world-spanning consequences and complexity. This means that we can also confront the idea of what the systems theorist Philip Garnett calls total systematic failure, in which 'there is no outside; all global systems are connected, and we are in them' (Garnett 2018). It is at this nexus of information and power that Anthropocene conditions are made clear as the material entanglements of modernity meet.

The 2008 financial crisis and its fallout is another example of these processes, as the interdependency of ostensibly national and individually regulated financial markets was made visible through its almost total collapse. It was only with the express intervention of governments and central banks that the complete disintegration of the economy as we understand it was avoided. In its aftermath, however, this system failure was explained away in the United Kingdom as the result of spending largesse and fiscal irresponsibility (Berry 2019), meaning that media consumers were not made fully aware of the way their own lives were fully intertwined with those of people on the other side of the world, or how the collapse of the Icelandic banking system (population 356,000) could trigger defaults across Europe and beyond, or how this system had been built up in the first place and whether or not they had given any consent for their capital to be used in such a way.

Charismatic data

Headlines that promise news of 'how A is changing B' or 'why Z is changing the way we live' often attribute agency to specific things as if they are immaculately conceived before beginning to show effects – a great man theory of history tagged to specific events or technologies. In reality, these happenings are part of chains of consequence which were set in motion long ago and that will continue to have impacts long into the future. This is what an Anthropocene perspective brings to individual events in the news cycle, and indeed to our understanding of the idea of discrete processes more generally. Anthropocene data journalism is fundamentally about narrating the central story of the epoch through its disparate but interconnected macroprocesses, and not merely the spectre of climate change.

Modernity, then, with its own appetite for quantification and ordering, produces an informational shadow of environmental science, stock market fluctuations and productivity figures. The trajectory of the Anthropocene is visible not just as scientific data, but in the systems of data formation themselves that are churned out at an exponential rate. Yet in this river of information the foremost data set of the Anthropocene remains the one that tells the story we need most acutely to find a resolution to, namely the mapping of atmospheric carbon from the 1700s to the present day.

The Mauna Loa Observatory on Hawaii's daily measurements are ground zero for the Anthropocene's climate record. The observatory is 3,397 m above sea level and has continuous and reliable data to help plot the familiar gentle course of the 'Keeling curve', named for Charles Keeling's pioneering work to measure atmospheric CO_2 in the Pacific that has proven crucial to confirming greenhouse gas-induced climate change. The curve – along with its outriders marking a range of parallel processes in the dance between carbon, energy, movement and consumption – is familiar to most of us even if we do not understand the specifics of what it is saying, arching slowly upward as the 20th and 21st centuries have advanced and temperatures have risen.

Ultimately carbon and modernity go hand in hand, as does the explosion in information that the energy intensive data society has created. Plotting the Anthropocene does not have to be a grand undertaking. It can be something as banal as people finding that their shopping costs more than it used to because of the collapse of an infrastructural pathway, or that the second-hand car market has spiked due to the collapse of auto finance models. These multiple manifestations of data are part referential, part experiential, and in these moments of contact with everyday life the data becomes charismatic in the broader stories we tell about our own lives and the world at large. In understanding the relationships between different processes and making those explicit, we can perhaps move beyond data journalism as simply a tool and a new grammar for old problems, and instead find the charismatic links that deepen our grasp of the global present.

6

Reporting the Climate City

On a darkening November afternoon in 2021, crowds lined the streets of Glasgow and photographers stood on bridges over the motorway in the frigid air, all hoping to catch a glimpse of The Beast. The nickname for the US presidential limo carrying Joe Biden to a reception for world leaders at the city's castle-like Kelvingrove Museum had stuck in the popular imagination, to the point that The Beast with its darkened glass windows were a metonym for Biden himself. The Beast became a rolling spectacle of American greatness in a city infatuated with and intimately bound up with transatlantic exchange.

In the age of fossil culture, seeing the bombproof limo with its V8 engine snake through the closed streets of Scotland's largest city to applause and awe embodied the entanglement of energy, global power and brute force. The circus that had descended on Glasgow for the UN climate COP occupied the domestic media for weeks as the public lapped up stories of international politicians tucking into local food or buying soft drinks and coffee at convenience stores. Behind the circus was the creeping awareness that climate negotiations were something of truly historic importance, even if their inner workings and their implications were broadly opaque to the people outside of the museum's ring of steel.

The Scottish government, keen to position itself as a major voice in the new geopolitics of climate change, took the opportunity to angle the country as a sustainable powerhouse at the forefront of the push for a green modernity. The city council provided free electric buses to delegates as locals were forced to negotiate road closures and police lockdowns of their neighbourhoods with rooftop snipers. Visitors were given smartcards for trains and the subway that residents had been told were technically impossible by their elected representatives, and the local rental market spiked by 1,000 per cent in some neighbourhoods as landlords cashed in on climate fever.

Elsewhere in the city the frictions and contradictions inherent to the climate roadshow came to the surface. A group of indigenous activists from communities around the world, who had occupied a derelict building after they were unable to pay inflated market rates for accommodation, ended up in a stand-off with the combined UK police forces. Climate demonstrators were arrested and kettled, and the *New York Times* (with corporate sponsorship from investment banks) spent millions of dollars on a climate hub where for a few hundred pounds visitors could hear roundtables with strategic thinkers, youth activists and the *New York Times*' own star reporters on how to tackle the climate crisis. The whole world was in Glasgow, bringing colour to the greyness of winter in north-west Europe, but the presidents, lobbyists and activists who flooded into the city were all operating in a space already teeming with stories about the Anthropocene well before the COP rolled into town.

Cities are one of the defining features of the Anthropocene. As the urban studies academic Stephanie Pincetl describes it, urbanisation and global environmental change go hand in hand, and cities are one of the primary agents of the Anthropocene problematic, their growth perpetuating not just rises in carbon emissions but a desire for food, raw materials and technology that has risen in lockstep with the established Anthropocene markers (Pincetl 2017). The proportion of the global population living in cities has been rising steadily (alongside the increase in global population more generally) since the 1700s and it is now estimated that just below 60 per cent of all humanity is now urbanised.

The city has come to frame much of contemporary journalism, for better or worse – accusations of metropolitanism levelled at the media can be countered with the argument that that is where a large slice of the population of any given country is. London and its hinterland has a population almost two thirds higher than the entire nation of Scotland. It continues to suck in capital and people from the rest of the British Isles, and in the past 30 years has been transformed from a grimy and dilapidated former imperial capital into a modern globalised hub, with the help of huge public infrastructure spending, private finance and generous tax loopholes. Tellingly, the British media industries use the label 'nations and regions' to refer to the places beyond London, despite them constituting the majority of the population and area of the UK. This includes Glasgow, centre of the Clyde Valley conurbation and previously regarded as the British Empire's second city and industrial heartland.

London's dominance is by no means unique in this regard, sitting in the rankings of top global cities alongside New York, Shanghai, Beijing, Tokyo and Paris as one of the megahubs of contemporary modernity that dominate not only their own countries but the global economy as a whole. The rise of the global supercity has created an interest in urban and cities journalism that has also put questions of urban sustainability and public transit into the mainstream, speaking to audiences of lifestyle-conscious city dwellers and those rubbernecking the car crash of contemporary urban development more generally. Projects such as Citylab – now owned by the Bloomberg Group – and *The Guardian*'s Cities project brought these questions into sharper focus in the English-speaking world in particular, and around the globe there are innumerable examples of spin-offs and startups capitalising on the demand for journalism about cities and the broader developmental questions they pose.

As well as being a source of environmental problems, cities are crucial to moving past the current crises of human impact and form both challenge and solution, not least with regard to climate change. Although living in a city is far more efficient in terms of overall energy than living in strung-out rural communities or small towns, cities are still the largest sources of carbon emissions due to their sheer size and scale. Beijing is famous for its regular smogs, as are Los Angeles and the conurbations of Bangladesh, Pakistan and India. The Taiheiyō Belt in Japan runs for over a thousand kilometres and contains upwards of 70 million people, while the rest of the country suffers from depopulation. In the city we find both environmental crisis and dreams of its resolution, landscapes of prosperity and promise and nightmarish visions of a world without nature.

Glasgow is a city between eras, at least in the official narrative of post-industrial resurgence that underpins its contemporary politics. It is also a laboratory for understanding modernity's toxic legacy and for grasping the dissonance between climate reality and the narrative of economic development that is baked into how we view the urban space.

Glasgow is also (in)famous for its ability to show how cities can have peculiar environmental and sociological effects. In both social research and the popular press it has been widely documented that living in Glasgow is bad for you, a phenomenon that has come to be popularly known as the 'Glasgow Effect' (McCartney et al 2012; Walsh et al 2017). People in Glasgow have a lower life expectancy and general health than might be expected given their income distribution and education, and explanations have ranged from the effects of the local food system to intergenerational trauma and the stress of the urban fabric itself. Worldwide, Glasgow has attracted interest for its apparently mysterious ability to make people sicker just by being within its confines.

Despite having once been one of the major economic hubs of the developed world and a key component of British industrial imperialism, Glasgow today is a landscape of crumbling Victorian grandeur, cheap neoliberal development and social democratic regeneration of varying quality. The clamour to understand the Glasgow effect has made it an international talking point and a lesson in the perils of post-industrial economies and regional imbalance – a tapestry of never fully realised projects and attempts to modernise and renew. Glasgow also allows us to travel into the future to see how the megacities of the present might look once their periods of exponential growth tail off. In the early 1950s Glasgow was one of the most densely populated cities in the world, but through both economic decline and political planning it began to lose its population, leaving a hollowed out streetscape of empty lots, motorways and isolated neighbourhoods. Crucially, it has not built any new metro routes since the late 1800s, and has been relegated to the status of a hanger-on in the new constellation of globally significant supercities, despite its key role in the development of the world we live in today.

Glasgow's reputation as a bastion of science and industry is well known, but along with the rest of urban Scotland it was also a powerhouse of high modern journalism. Glasgow's status as a leading city of the Industrial Revolution gave rise to a flagship morning broadsheet of international standing, the *Glasgow Herald*, as well as an evening tabloid aimed at a more working-class readership. The city's journalistic capacity has undergone huge atrophy in the past two decades, but the old mastheads still exist as virtual publications run out of a single office. Like their Atlantic cousins in places like Chicago or Detroit, they were a key part of the fabric of the industrial past. The city's suburbs and satellites also enjoyed their own local papers, some of which still exist as brands but with centralised production operations and little local presence. Despite now being run from a modest office holding a dozen employees at a time, the *Glasgow Herald* still makes claims to be a quality newspaper of international standing, not least as a title that at 240 years old has been with the world from the steam engine to the quantum computer. Nevertheless, in Glasgow the news industry is as post-industrial as the manufacturing industries whose decline it has described in detail. As the capacity of the press has been diminished, so too has the ability of the city to understand itself.

What, then, can cities journalism do for a city like Glasgow and the way we narrate the role of cities in modernity and the Anthropocene? Like the devastated wastelands of New York and its artistic counterculture described by Marshall Berman in *All that is Solid Melts into Air*, Glasgow today is a whirl of artistic gentrification, private development and land speculation, and of governance structures caught between the needs of people and the siren cries of international capital and external investment. As a laboratory of the present and the future it gives us access to countless stories about possible outcomes for the world at large, both positive and less so.

Finding stories in the climate city

Walking around Glasgow today is a 500-year tour through the stages of urban growth, mercantile and industrial capitalism, and decline. From the near-demolished remnants of the medieval city to the shopping centres and IKEA showrooms of the former docklands that sprang up on vacant land in the 1990s and 2000s, Glasgow is a city of immense and constant change. On the south bank of the River Clyde the international financial group Barclays have constructed a new campus that looms over the empty squares of a former industrial district, taking advantage of cheap property prices to farm out aspects of its London-based businesses and save on labour costs. Nearby new luxury apartments rise up to house these finance workers, and the private developer is paying for the creation of a new waterfront boulevard as a hat-tip to sustainability. Some in Glasgow are going to grow very rich off this regeneration, others will not, but it is also just the latest iteration of Glasgow's centuries-long dance with capital.

The situation just downriver from the finance campus in the Thornwood district is a case in point. A traditional neighbourhood composed of Glasgow's signature red tenement housing, in the 1960s the local railway line was closed and replaced by an expressway that runs on concrete viaducts through the heart of the area, skimming the backyards of those closest by. Residents now live with the constant rumble of cars and the impacts of pollution on the local streets, and the former harbour area is home to drive-through chain restaurants, car dealerships and luxury apartments with ample car parks.

This is a point of contention in the local community, with battle lines being drawn between different visions of prosperity (Hinde 2021). The city of Glasgow as a whole has the lowest proportion of car ownership of any city in Britain, yet it also has some of the worst public transport, and has been extensively redesigned around petromobility. Glasgow is often a city at odds with logic, but in its failures and oddness it creates space for semi-utopian futures too. In a place where the foundations of the contemporary settlement are so obviously precarious, it is easier to suggest ways out.

Climate change is already fundamentally changing the way cities function, and city planners are responding accordingly, on paper at least. A 2.6 degree global temperature rise in line with current data produced by the IPCC will have severe implications for cities in tropical and subtropical climates, with cities in temperate and subarctic climates such as Glasgow potentially taking on a role as climate refuges[1] for those fleeing natural disasters and extreme temperatures.

Glasgow is no stranger to refugees, with a significant refugee population attracted by its existing diaspora communities and the UK government policy of placing refugees and people claiming asylum in the city as part of their dispersion policy. With worsening climate change, however, the number of refugees seeking safety in the British Isles is likely to increase due to both direct climate effects and conflict as a result of climate. Glasgow will also suffer from significant climate impacts itself, not least the possibility of rising sea levels and increased excessive rainfall in a city already heavily defined by an extremely wet Atlantic climate. This means that although Glasgow may be able to function as a climate refuge compared to other more vulnerable cities around the world, it will still have to contend with regular flash flooding and tidal surges, as well as damage to its broader infrastructure from extreme weather. Wherever we look we can find spaces where climate is already being felt or will be in future.

Like every other major city, Glasgow is faced with a very particular future – alongside reducing carbon it also has to be climate-ready, and ready for changes not just to its infrastructure but to its social make-up as well. This goes far beyond longstanding interests in green spaces, cycle paths and the traditional concerns of urban greening. Through Glasgow we can think about different approaches to the future of the city as an idea, with clear climate objectives that speak to larger Anthropocene experiences.

Roads to the future

When Glasgow's express road network was constructed in the 1960s, a huge swathe of the city was destroyed to build trenches and viaducts for the new M8 motorway. This effectively split the city in two, separating the wealthy west from the centre, but also the poorer northern suburbs from its central districts, making it extremely hard to walk through the cityscape and to move freely and safely without a car. This was worsened further by the destruction of the municipal tramway system in the decade before and the privatisation of the bus system as part of the financialisation and deregulation of the 1980s in the name of modernisation. Glasgow illustrates well Jane Jacobs' (2007) observation that it is the automobile, and not drugs or television, that has done the greatest damage to people's wellbeing since the 1970s.

The lifetime of this concrete infrastructure has now expired and needs expensive temporary repair and eventually replacement if the motorway is to continue functioning. What was once the future now represents an expensive and dysfunctional legacy, a reminder that modernity has blind alleyways as often as it has shining paths to progress. Campaigners have seen this as an opportunity to reconnect the different parts of the city, but also to turn the motorway corridor into a linear park, which would remove carbon from the air, reduce overall city temperatures during heatwaves and function as a water sink for flash flooding. Less radical plans include reducing the number of exits and covering the area to turn it into a park at the edge of the city centre, hiding the scar in the cityscape that has divided Glasgow into disparate parts for two generations.

The questions and stories that arise here are not just about the future of the motorway – and there remains a powerful road lobby in Glasgow, as in every major city – but about the fabric of the city itself and how these interventions removing or diminishing the presence of the motorway could be of benefit to the city and its people. The future of the road and its alternatives are also a choice about wider cohesion, and ultimately about the success of Glasgow as it seeks to both adapt to climate change and flourish economically and socially.

Glasgow's rolling acres of concrete and tarmac are at the centre of a debate about how society as a whole should work – those opposed to closing the motorway claim that it would damage businesses or make life harder for 'hardworking families', as the emptying out of the city has created new US-style car-centric suburbs. The campaign against the motorway is also interpreted as a wider attack on a way of life that has come to define much of local politics, of individualised wealth versus collective infrastructure and a post-industrial shift from manufacturing communities to retail and service jobs, speculation and a new lower middle class on the urban fringe. The infrastructure of this system is made up of slip roads and shopping parks, drive-through coffee franchises and credit-fuelled car ownership that claims to liberate but fosters reliance. The motorway is not just about transport, but is a visible sign of how the city sees its economic and social priorities, bleeding into everyday life.

The alternatives to the motorway also offer the opportunity to explore completely new systems or organisation for the climate city. Walkable neighbourhoods, repopulating the city centre with people, building schools and community facilities, and linking the parts together with a new metro system all float tantalisingly in the media air as better uses of the tax money earmarked for maintaining the crumbling concrete. More than anything else perhaps, as other cities around the world experiment with motorway removal and the end of cars in the urban core, Glasgow looks increasingly anachronistic. The motorway story is the story of a city out of time and out of step with the present.

Feeding the city

Climate change constitutes a significant threat to regular and adequate food supplies even in developed countries such as Scotland, a foretaste of which was given in 2022 by the food shortages and economic shocks arising from the war in Ukraine. The COVID-19 pandemic too showed the issues with world-spanning food networks and a reliance on transnational flows, something which in the UK has been exacerbated by the disintegration of trade policy and difficulties with managing the complex infrastructure needed to get food to where it needs to be.

With food, as with so much else, Glasgow is a place of ironies. It has had problems with malnutrition and access to food but is also home to radical innovation in local food networks, while Scotland as a whole attempts to market itself as a food destination with billions of pounds worth of exports each year. Delegates at the Glasgow climate COP were bombarded with information on Scotland's sustainable food culture and Joe Biden and the other heads of state dined on the finest Scottish produce inside the summit's ring of steel, but like the electric buses and electronic travel passes this was a far cry from the everyday experience of the average Glaswegian.

Glasgow's vast swathes of derelict land have sat for decades waiting for the day they become valuable enough to develop. Even in the richer districts in the west of the city, empty space is left without being cultivated or used, but Glasgow has also seen a small and growing urban farming movement develop. One project run as a nonprofit has both established markets for and acquired sites to grow locally produced food. Using three microfarms within ten miles of Glasgow, one of which is actually within the limits of the city itself, food is produced for the project's member networks and combined with other local sources of dairy and environmentally sensitive products to provide a package for home delivery or for retail through the project's stores.

The use of microfarming and urban farming, generating food close to where it is consumed and reducing reliance on complex networks and infrastructures, means that the food supply is less subject to large-scale disruption. The project reinvests its profits and operates a credit-investment scheme for members to expand and add further assets and farming sites, with ambitions to develop a regional and national network of low-carbon and resilient food production and distribution. Not only does this mean consumers buying more sustainably, but also control of their own food supply and a degree of democracy in how their food is provided.

Areas used for urban food production also provide a net benefit for the area around in terms of groundwater accumulation and biodiversity, and when farming land replaces tarmac, concrete and gravel, significant reductions in urban surface temperature as well. Everyday stories about food and the rituals of eating reach into far deeper questions of sustainability and economic deprivation, but they also challenge the binary between the city and the agrarian periphery, creating new kinds of politics and challenging core narratives about the city and the countryside.

Are a few small farm projects in a city of over a million people a viable solution? They rank alongside the community news organisations and local bloggers challenging the city's legacy media in terms of impact and scale, but they work, and they show that the city can change quickly and effectively. Even more, perhaps, they show that new food economies are possible, and they provide stories to tell about what these economies look like.

Anthropocene homes

Like many cities around the world, and despite a surfeit of cheap land, Glasgow has experienced increased housing costs relative to income over the past four decades. The decline of municipal housing and a reliance on private developers has resulted in a patchwork of buildings of varying quality dominated by private landlords and speculative investors, and no clear mechanism for bringing the thousands of historic buildings that the city still has up to date with the demands of climate-safe engineering. The city is in desperate need of improvements to its housing to cut energy use and increase resilience to climatic shocks, but also for reasons of safety and welfare as people struggle with energy costs and basic issues like damp, outdated electrical supplies and disintegrating buildings. Glasgow is one of the wettest cities in Europe and the climate takes its toll on old buildings and newbuilds alike, with many older buildings merely being demolished altogether once the wind and rain have been left unchecked for a few years.

The housing crisis provides an avenue to explore Anthropocene issues in a number of ways, from questions of fuel poverty to resilience against extreme heat, but also the need to massively upgrade and retrofit housing at huge expense and in a short space of time. Irrespective of climate change, the need for intervention in the housing sector has been a headline-grabbing issue worldwide, and so the need for climate resilient low-carbon housing feeds into more general concerns about quality and provision of homes for people. There are obvious tie-ups between the use of low-carbon housing to reduce emissions, the construction of housing that is itself climate resilient and able to protect people from extreme climate effects, and the larger impacts of an unrestrained market on the ability of cities and states to deliver public goods in the common interest.

There are plans to retrofit Glasgow's historical housing stock, but this is also extremely expensive to do and requires large amounts of public money or state-backed debt finance. This gives rise to a political conversation about who should carry the cost of adapting to climate change. The estimated cost of retrofitting Glasgow's housing is around £10 bn according to the city council's own estimations, representing a huge investment in public goods that is out of step with current political culture.

As with the problematisation of highway expansion and the maintenance of infrastructures which are not fit for purpose, the housing crisis is an opportunity to get a narrative handle on the sheer scale of the very basic problem of how we live. Housing is the gateway to an entire conversation about how we manage space, buildings and assets in the public and environmental interest, but also how we view the idea of property itself in a time of crisis. In a city that for decades has been carved up by speculators and retailers demanding cheap land and plentiful car access, the end of this settlement completely changes the political economy of the city and forms the basis of new economies altogether.

Green jobs and sustainable industry

Where these economies will come from is the question at the heart of Glasgow's future. As a post-industrial city Glasgow has struggled with the impact of deindustrialisation on its economy and identity, reaching a population nadir in the early 2000s and showing only modest recovery since. As the historically largest manufacturing hub in a country with a growing renewables industry, there is little sign of this on the ground. Scotland as a whole is still reliant on foreign manufacturing and there has been much discussion of the failure to preserve skills and jobs in manufacturing and heavy engineering. Despite having some of the best preconditions for renewable energy production in the world, Scotland has generally failed to harness these developments as an engine for larger scale social change or decarbonisation (Gibbs 2021).

Transition to a climate city offers opportunities to engage with everyday questions of labour and employment, not least in the creation of quality jobs and the use of low-carbon and zero carbon technologies, and to redraw the map of the post-industrial cityscape. The need to rethink infrastructure under climate change means that novel technologies will be necessary to adapt and that existing modes of production will no longer be suitable – particularly alternatives to carbon-intensive processes such as concrete in buildings and streetscapes, and the need to produce steel without producing carbon using hydrogen, which can be readily produced from renewable energy at scale given Scotland's renewable reserves. All of these possibilities exist in a web of regulation, investment, innovation and political will, with everyday implications in the form of jobs, health and wellbeing.

Green economies are not just questions of decarbonisation and climate tech either. Post-material service economies and circular economies create new forms of social life unfamiliar to the current settlement. In 2022 Scotland adopted a much-vaunted roadmap to a circular economy, but interrogating what this actually means is far tougher. There are material circular economies of reuse and recycling, but also social circular economies in which value is not extracted but flows back into the communities that create it. Profits need not flow to tax havens or head offices halfway around the world, but into transition itself. The green economy can either acquiesce to the transnational processes of the Anthropocene with some low carbon quality assurance, or it can reconfigure them to deal with globalisation's 'bads'.

Glasgow's economy represents both a relic of the fossil age and a future imaginary built off the back of its negative experiences of deindustrialisation. The chance to create meaningful jobs and economic security goes hand in hand with technological adaptation, and the development or absence of these green industries provides a backdrop to the future development of the city and the realisation of a post-carbon economy. The outcomes may be utopian, but fundamentally these are bread and butter stories of everyday life.

Networked narratives

Each of the topics mentioned bleed into the fabric of everyday life, and in the moment of transition between the contemporary city and the climate city each one offers an opportunity for journalistic intervention. None of them need to be characterised as environmental stories, but all offer ways into talking about important issues that impact on the future of the city as a whole and the wellbeing of its people.

Beyond these location-specific issues, cities are also characterised by their dense networks and scale, which reach to the heart of the Anthropocene problematic. Sociologists saw modernity's early stages unfold in the bustling chaos and new social relations of Vienna, New York, Manchester and Chicago, but bearing in mind that the Anthropocene is visible by its networks as much as concentrations we can also begin to rethink how we see the space beyond cities and their inclusion in the overall structures and conditions of the Anthropocene. Cities experience a form of quantum entanglement with one another and other ostensibly unconnected places. Not only are cities highly reliant on their hinterland for their food and energy needs (the Highlands of Scotland, for example, generates 327 per cent of its own renewable energy demand[2] and exports the rest to the lowlands and cities), cities often have enduring links into other spaces and places due to internal and international movement. Migrants, rather than being an addition to a city, are in fact crucial to the constitution of cities: what David Morley (2017: 7) calls places where the modernities of the future are already taking shape.

Glasgow's hinterland exposes different networks of human movement and interaction, which as Ash Amin and Nigel Thrift (2016) for example have argued, is crucial to seeing a city as more than delineated space. To the north and west of Glasgow the population density drops considerably, but the hinterland of Glasgow extends all the way up Scotland's west coast where people rely on the city for services. If you have a heart attack on the Isle of Barra, 50 miles out into the Atlantic, you will be flown by helicopter to a hospital in Glasgow. If you need to go shopping, a five-hour drive into Glasgow is the norm for many people, but while these rural communities rely on the city, the city also relies on the rural communities. Wide tracts of semi-wild land that provides important ecosystem services, leisure space and acts as a carbon sink for the city's emissions are crucial to Glasgow's existence. Beyond this there are also transnational networks a city feeds into, a globalised hinterland of extraction and consumption, and of human connecting. The various diaspora populations in Glasgow reach into more climatically vulnerable regions including Pakistan, Ethiopia, Somalia, Nigeria and China. These are connections that are likely to be significant in terms of climate migration, but also in terms of connecting the city into the wider world and giving it a bottom-up rather than top-down cosmopolitanism.

Nor are all cities the same in this regard. Some like Stockholm or Paris have clear centres and peripheries; others are merely collections of neighbourhoods without any clear point of reference. Others have their own colonies and detached dependencies, offshoots that at first glance we do not even see as one and the same. One of the busiest air routes in the world is between Seoul in South Korea and Jeju, a moderate sized island off the country's south coast. This allows Jeju to function as Seoul's beach, with flights between the two departing as frequently as a local train might leave between two geographically adjacent places.

In a less extreme example, London recently spent billions of pounds on a cross-city express rail project that connects its main international airport directly to its business and financial centres, and on to the wealthy suburban conurbations that have grown up around the city. This system exists only to overcome the inconvenience of previous stages of development, dug deeper than any existing transit line and rising and falling through an underground maze of infrastructure, co-financed by government and the financial sector. Such schemes are built to meet an ostensible need, but they also beget new networks and interactions. Suddenly an unloved suburb on the Thames estuary is 18 minutes from the City of London and the centres of international finance. In Glasgow, meanwhile, where there is no clamour for wealthy commuters to move around quickly and no centre of capital, the rich drive or take taxis and the poor rely on patchy privately run buses.

How we experience cities is dependent on what their networks are and how they ultimately cause us to behave. More than anything else, looking at infrastructure helps us to see how cities are the result of the decisions we make. In developmental questions we often have a huge amount of agency that we choose not to recognise, and infrastructure is often portrayed as a deterministic unfolding rather than a series of specific choices to which we have alternatives. One of the challenges of reporting on these networks is to break down the structures of power behind them, and to find opportunities for agency in the cracks.

New journalism for Anthropocene cities

The new journalism of the city is not just the desire, then, to tap into the booming market for urban stories, but to understand how the everyday narration of urban space contains the key to more productive futures. Cities are and will increasingly become prime sites of environmental shock and change in the coming decades, in addition to the impacts they have already had on the climate and local ecosystems. Images of the flooding of the New York subway or of huge mudslides in China show the results of bad planning and historical vulnerability, but they are also focal points for other Anthropocene problems beyond extreme weather events. Journalism can take us to these extreme places but it can also tell us more about our own backyards, both metaphorically and literally.

The unprecedented growth of cities is itself an Anthropocene marker, as these cities increase surface temperatures with their concrete and asphalt heat reservoirs, interrupt water systems and of course produce disproportionate amounts of greenhouse gases and pollutants. They also have other long-term impacts on human health and can negatively impact on human wellbeing through both increased exposure to toxins and stress, and economic strain, alienation and constant revolution.

In Glasgow, the Glasgow effect and its multifaceted causes illustrates well the bind between environment, economy, urban space, the past and the future. Well beyond Glasgow, though, each city has its own stories to tell about the drivers of the Anthropocene, of capital and modernity and its inequalities, its winners and its discontents. If we are to move beyond the Anthropocene and rid ourselves of its markers then cities will have to fundamentally alter themselves too, in ways that we currently find hard to even conceive of. That entails seeing the city as both a banal everyday canvas and a dreamscape.

Cities journalism is not merely a way of documenting life in the contemporary city, as important to journalism as that may be and where concentrated markets of media consumers allow it, but of drawing the connections between cities and the wider world. This means stressing how cities, in both their toxicity and their vulnerability, can be made to meet the need for drastic reductions in their footprints, and to function as places of security and stability on a rapidly shifting planet.

7

New Times for Journalism

In Kurt Vonnegut's antiwar novel *Slaughterhouse-Five* the autofictive narrator speaks to a filmmaker about the futility of writing pacifist literature. 'Why don't you write an antiglacier book instead?' the film producer asks Vonnegut's protagonist, reasoning that human conflict is as inevitable and impossible to hinder as the slow march of glaciers towards the sea.

In light of the invasion of Ukraine and the destruction of Gaza, the horror of the carpet bombing of Dresden and its cultural memory find a renewed resonance, yet the most conceptually dissonant thing about Vonnegut's glib exchange is that glaciers no longer slowly march towards the sea in the same way they did in the middle of the 20th century. Humanity has become highly adept at stopping glaciers, be they in Iceland, the Swiss Alps, the Himalayas, Patagonia, Greenland or Antarctica. Yesterday's truisms come back to haunt us with irony, and even recent history can seem decidedly alien. Time and events have gone awry, and we are still catching up.

Unsurprisingly, the state of the world's ice caps forms a red thread through decades of climate change reporting, focusing on both the science of collapsing ice sheets and the strangeness of Antarctica and the High North as a front in the war to prevent catastrophe. A story beloved of people with an eye for the spectacular and weird is how, in the 1950s, the US military established missile bases buried in the Greenland ice cap, treating both war and the immovability of the vast glaciers covering the world's largest island as facts of life. The planners of what was christened Project Iceworm imagined the ice as a solid constant, but they fought a losing battle against the shifting frozen caverns they had created, whose timescales they were forced to appreciate. Even so, when the bases were abandoned, machines, rubbish and nuclear material were merely left in the ice, in what was assumed to be a perpetual frozen wasteland that would slowly be buried by fresh snow and crushed by the weight of the glaciers pushing down on it. At some point it was thought the nuclear material and bent mental would be crushed into bedrock and vanish into the earth.

Today we know all too well that glaciers are not inevitable or impervious to change, and that in the Anthropocene ice caps are transient phenomena which may vanish within lifetimes. Nuclear waste, on the other hand, lasts longer than we can easily conceive of, created in a few short years but dangerous for millennia.

Time, then, is not what we thought it was. How, in the space of half a century, can our concept of time change so much, and what does this mean for how journalists deal with time? Just as creationists struggle to comprehend the age of the earth through their anthropocentric world view, in which the earth and humanity temporally go hand in hand, we are today faced with the opposite problem. We cannot easily understand how deep time has become so liquid in front of our eyes.

As the ecocritic David Farrier (2020) notes, uranium, older than the earth itself and forged in the heart of stars, has since 1945 become a malleable tool in a short flash of human history. Thanks to nuclear energy and warfare, plutonium-239 is now scattered around the world and stored in subterranean bunkers, a particularly deadly form of human waste which puts humans at the beginning of another cycle of deep time. Plutonium-239 has a radioactive half-life of 24,000 years, meaning it remains a danger to humans, animals and plants alike for longer than anything resembling human society has existed.

Once it became clear that nuclear waste would be a long-term issue well beyond the present, the US government established the Human Interference Task Force to come up with a solution that could keep nuclear waste and people separated and secure for 50,000 millennia. The taskforce was given the job of communicating across deep time the fact that nuclear waste was there and remained dangerous, in forms of media that nobody had ever before considered. The solutions proposed were manifold, from a growing field of signs in increasingly modern languages to the encoding of warning signs in the DNA of local plant life, and these were merely some of the less bizarre proposals. The philosopher Stefan Skrimshire (2018: 314) describes these warnings signs as attempts 'to communicate something not of our present (why we need this waste) but of a far future fact that it will still be there and what to do about *that*'. The Human Interference Task Force was dealing specifically with the Yucca Mountain Nuclear Waste Repository in the USA, but this was about more than mounting a 'keep out' sign. In the space of a few hundred years European colonialism had come in and destroyed the indigenous culture of Nevada, serving as a reminder of the short half-life of languages and cultural forms more generally. This becomes a problem when considering that the oldest spoken languages on the planet with a common thread to their ancient forms, such as Greek and Tamil, range between 3,000 and 5,000 years old. There is still contention about how long exactly waste should be stored for, but the difference between half a million and a million years seems almost academic in the circumstances, as nuclear modernity's long shadows stretch off into the future.

Writing memory

Like project Iceworm and the Greenlandic military bases, the nuclear waste dilemma is a captivating story, but more importantly it forces us to think expansively about the nature of the future and our relationship to it. Just one of the many ways in which the Anthropocene plays with our notions of time, it demonstrates the extreme difficulty, if not the impossibility, of producing media which can bridge these barriers. It also exposes ironies in how we deal with the bads we create without even realising. Nuclear power may well be critical to stopping runaway climate change and destruction of natural systems across the globe, yet it also ties us into hundreds of thousands of years of careful guardianship, remembering where we have buried waste and why. As things we thought were constants change before our eyes, the products of a few decades linger with us for hundreds of thousands of years. Such are the ironies of Anthropocene time.

The exponential Anthropocene markers shown in the hockey sticks of the Great Acceleration all leave traces for the far future, from microchips in landfill to soil contamination and decimated forests. Plastics and heavy metals from products with a lifespan from five minutes to a few years can sit in the ecosystem for centuries before being broken down, continuing to impact on the world around. A question to ask for any object that is part of an event or news happening is not just 'what is this doing?' but 'what will this be doing for the next few thousand years?'

Time is tricky when it comes to journalism, and journalism is not always adept at breaking out of its temporal norms. This is not only because surveying the past can be heavily editorialised, but also because the future remains so malleable and the news cycle so glib about long-term concerns. As the communications scholar Carolyn Kitch has noted, journalism engages in the construction of 'useful memory' in that it is able to build (and rebuild) the past by constructing neat and often simplistic chronologies of global history (Kitch 2006). The immateriality and quickness of media also means that the futures it creates can appear inconsequential or changeable. Journalism likes to speculate, and will then happily change tack once a new or better angle comes along. So, while journalism is critical to building the future and the past, it is far from consistent in its approach, and its status as a practice of the extreme present means that it can often fail to connect backwards and forwards in the ways that are necessary.

Understanding memory is important because it helps to explain how and why things are the way they are and how they will come to be. The pushback against 'rewriting' of history as part of ongoing culture wars across the globe objects to the deconstruction of the simplistic and unproblematised narratives of colonialism and capitalist development of the past four centuries, and to the way in which contemporary conditions are related to these histories. Extractive capitalist settler colonialism in both its older and more contemporary forms has a huge amount to answer for in terms of Anthropocene problems. The connections with the past which are sometimes talked about in terms of cosmopolitan pride, such as British engineering in the Brazilian interior or the colonial place names that pockmark the Indian railway map, involve a rewriting of these Anthropocene histories in the contemporary language of globalised interdependency and benign cosmopolitanism. More than anything else, debates about empire and historical culpability reinforce the subjectivity of global memory and how it can trap us in particular stories about the present.

Deep time and crisis time

The media has no shortage of stories about our encounters with the temporal connections we have created in the Anthropocene. Anxieties about nuclear war have been replaced with the aforementioned anxieties about nuclear waste storage from both weapons and energy, or about the climate crisis which has slower but similar impacts on human wellbeing. Time is long and short at the same time. Some of these temporal changes are easy to conceptualise – urban sprawl, from the ad hoc slums of developing countries to the pop-up plywood suburbia of the American west, is shocking on human timescales. There are people alive who remember when their local shopping park was a meadow or forest, or when their city was just lines on a map with notional boundaries. Something as basic as a campaign against a new road or resistance to turning a natural river into a covered channel trade on our awareness of what has been, but also what might be.

Indeed, the enthusiasm for rewilding sweeping the Global North sometimes seems based as much on shared memory of lost childhoods and Ruritanian dreamscapes as sound ecological research. Conversely, the motorists campaigning against the introduction of cycle lanes, low traffic neighbourhoods and pop-up parks live in a state of certainty which is only a few decades old. The contemporary habit of car ownership and vehicle access which is normalised by the discourse of the public media is by no means as deeply historically embedded as we are sometimes led to believe or as traditional as we remember.

Other temporal transformations are much harder to confront or understand; our great-grandchildren will be as far removed from us in time as we are from the children of the 1880s, but they will no doubt continue to live with the legacy of climate change that we have created, and may in fact have a lower standard of living than many of us enjoy today as a result. The short cycles of the present have long-term implications.

The geographer Mike Gasher writes about journalism as a form of cartography (Gasher 2015), in which journalism itself functions as a web of mapmaking in the public imagination. Gasher describes reporting and mapmaking as parallel functions, noting that they do not mirror the world but are representational, with discrete borders and specific typologies and tools. Both mapmaking and journalism use a common language to render reality with an instrumental purpose: to construct the boundaries of what is acceptable and feasible. Journalistic cartography ultimately lends permanence to transient situations and can trap us in false understandings of the timescales we inhabit.

Contemporary thinking on the Anthropocene has also introduced the concept of deep time to more mainstream conversations, in which, in the words of materialist media historian Jussi Parikka, 'we are dealing with issues of time scales that are not necessarily authored only by the loose category of humans. Instead, we find ourselves orienting in non-human long durations' (Parikka 2016: 281). The geological time which has been used to divide the history of the earth into epochs is in the Anthropocene transformed into human time. What we do today creates future geology as these materials are crushed by the weight and pressure of the earth around them and transformed. These million-year timescales are far beyond the scope of most media approaches.

What then emerges are two forms of time: crisis time, which deals with the existential threats of the present, and deep time, which invites us to consider the deep past and the far future. These things are not, however, separate. Critics of the Anthropocene as a concept have argued that discussions of deep time and our small place within it reduce contemporary crisis time to a historical footnote, but this is a historical footnote of huge consequence.

Nothing perhaps better illustrates the collision of deep and crisis time as the 'doomsday clock' managed by the Bulletin of the Atomic Scientists, a rhetorical tool to showcase the severity of the global existential risks humanity faces. Started by veterans of the Manhattan Project that produced the atomic bombs dropped on Hiroshima and Nagasaki, the Bulletin regularly updates the 'clock' to reflect developments of critical importance to the future of humanity. In January 2020 the Bulletin symbolically moved the hands of the clock the closest to midnight it had ever been at that point, outstripping 1953 (the advent of hydrogen bombs) and 2018 (the failure of climate and nuclear nonproliferation talks). It explained its decision thus:

> Humanity continues to face two simultaneous existential dangers – nuclear war and climate change – that are compounded by a threat multiplier, cyber-enabled information warfare, that undercuts society's

ability to respond. The international security situation is dire, not just because these threats exist, but because world leaders have allowed the international political infrastructure for managing them to erode. (Bulletin of the Atomic Scientists 2020)

Since then, the hands have moved again to reflect the dangers posed by international conflict and Russian threats to use nuclear weapons on the battlefield. The Bulletin's updates are a semiregular feature of the news cycle, to the point of being a recurring curiosity in the science section of news sites, but the implications of their warnings are harder to make into news. The temporally wicked nature of climate change means that greenhouse gases are not a problem that can be immediately ended or managed on the timescales of government budgets and central bank investment packages. Unlike nuclear war, where the threat is of a main event that could be over in a matter of hours, climate threats are slow and spread out. This inertia operates over decades and the climate effects we are experiencing in the first decades of the 21st century are part of complex cycles and historical build-up. Even if the world can reach net zero in the coming decades, something that seems highly unlikely at the time of writing, this does not mean an end to climate disruption over the same time period.

One of the main tools the IPCC uses to guide climate negotiations is a series of emissions pathways – predictive outcomes based on different scenarios for emissions reduction with consequent impacts on global temperature. These emissions pathways are attempts to look into the future and cross some of the obstacles and barriers of contemporary political time. Every single one of them envisages levels of climate change which will fundamentally impact on life on earth, from the disruptive to the catastrophic, but what this actually means is harder to tell stories about. In forming time, what are the material political, social and economic processes that turn the high emissions scenario at the top of the graph into the low emissions scenario at the bottom (Figure 7.1)?

Journalism operates on timescales of weeks, and occasionally months. Because of the nature of news events, journalism must conform to specific news values (see Harcup and O'Neill 2017) which position it within a number of expectations about what these events mean and how they can come to impact on the immediate future. Journalism will often look forward to events which inhabit this future – major sporting events, scheduled elections, political summits such as the United Nations COP negotiations, or occasionally somewhat more far-sighted horizons such as the Millennium Development Goals, which were designed to transform and more importantly quantify global development. Such future markers help to structure the future and provide some temporal organisation to a practice which exists largely in the present.

Figure 7.1: Illustrative emissions pathways generated using data from IPCC AR5, showing the attempts to generate futures through IPCC models. Only the lowest line keeps temperatures within acceptable limits.

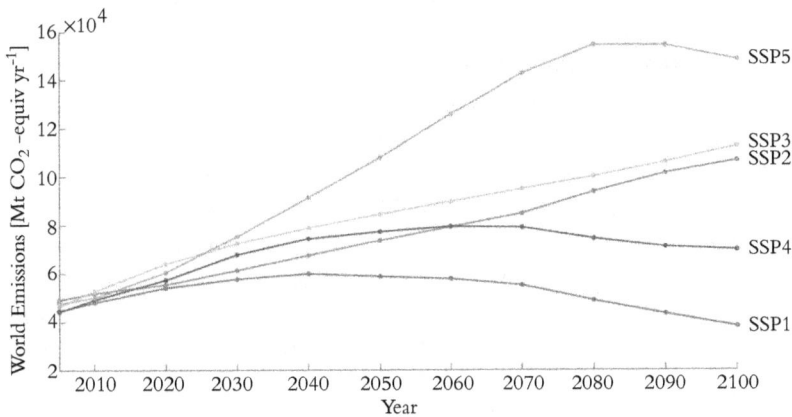

In an Anthropocene framework these time references of expectation have to be rewritten or expanded. The challenge is to expand the timescales of journalism beyond what we customarily use but, being mindful of critiques of deep time, to avoid becoming lost in the scale of the grand narrative with its semireligious overtones. Given that our agency is manifested predominantly in the here and now, the solutions to humanity's predicament must also be located in the reasonable frameworks of the present. This is one reason why climate change discussions often use benchmarks such as 2030, 2050 and 'the end of the century' both when making forecasts and engaging with the public.

The more immediate concern is that humanity needs to do something of unprecedented radicalism within the next eight to ten years to have a real chance of halting catastrophic climate change. This means that the solutions to climate change and other Anthropocene processes *are* communicable on the timescales of contemporary media, but in the context of the deep time and delayed feedback we know is inherent to the climate system.

Whose time? Audiences and publics

Time is an issue not just because we are running out of it, but because we do not all see time the same way. As the historian E.P. Thompson argued in his analysis of time discipline and industrial modernity (Thompson 1967), our position in societal hierarchies and economic conditions can change how we experience time and place under the structures of modernity, and journalism is part and parcel of this temporal stratification. Audiences for journalism cannot be seen as uniform, and neither should their concept of the future be. The anthropologist Kevin Donovan has discussed how in the digital information economy the relationship between time and economic status traps the more precarious within specific horizons and shorter term loops of so-called 'zero balance' lives (Donovan 2020), as opposed to the longer term investment cycles of finance and emotional labour of the more wealthy. This is one of the reasons why climate change – with its warnings of apocalyptic floods and wildfires – is click-friendly in a very superficial way, but the everyday politics of survival mean that spectacle does not always lead to action. Indeed, as Catherine Happer (2024) points out in her work on the changing publics of digital space, one of the impacts of disintegrating institutions of news has been the removal of the monopoly on the timescapes and ideological scaffolding of legacy news media. Happer argues that media and public previously engaged in a 'social contract' based on intent, however truthful or untruthful the media may have been. But once this monopoly on time and the world system unravels, things become more complicated.

This is also where Lauren Berlant's concept of cruel optimism once again resonates. The timescapes of aspiration and hope used by working people are often chimeric, and despite the unfolding horror of climate shocks in the media we nonetheless labour under the impression that the impossible is still achievable. As the climate forecasts suggest that the idea of the 'good life' aspired to by working-class people and those trying to move from subsistence to accumulation cannot possibly come to pass in the way that is promised, audiences simultaneously carry the weight of everyday survival and work and the promise of future improvement. In media research, Barbie Zelizer (2022) in particular has argued that journalism scholars do not engage enough with the temporal implications of journalism, and this can be taken to encompass the temporality of the lives of audiences as much as the speed of journalism itself.

The future looks very different to an upper middle-class family living in Munich, with multiple assets and clear plans for life advancement, than it does to a gig worker in Colombia without any social insurance who is compelled to work in order to stand still. Moving forward through time rather than time happening around you is the privilege of the few. Thinking about the distant future is not an everyday occupation, yet it is something that impacts on all of us, and conversely it is also something on which we all have an impact. This is mirrored in the conversations in journalism about 'fast' and 'slow' practice, in which judgements are made on the relative cultural standing of different speeds of journalism. In the world of the audience, time is money.

Given journalism's function as a practice of the extreme present, it is tempting to construct the Anthropocene itself as a facet of a polycrisis which encompasses challenges to democracy, development and other near-term problems, and one which quite simply attempts to reframe longstanding macroproblems as linked to specific short-term ones. While it is not unreasonable to seek proof of longer term trends in the contemporary news cycle and the need for news hooks, the hunt for something relatable manifests in this bundling of short- and long-term problems in ways that do not always allow us to understand the potential solutions.

As Donna Haraway summarises it, the Anthropocene is a period of severe discontinuities in which 'what comes after will not be like what came before' (Haraway 2015). Haraway's conclusion is that, located as we are in the present, the job of contemporary society is to make the Anthropocene as short as possible and from that point cultivate 'imaginable epochs to come' that can replenish what has been lost. The timescales on which the Anthropocene operates are hard to establish and convey precisely because it operates on long and short timescales simultaneously, and as each second passes in the present the outcome and scale of that future changes.

Likewise, as Barbara Adam (1998: 7) laments, the problem is that popular social theory is made by journalists, not by sociologists. While sociologists habitually try to see structure and nuance, journalists are caught in a bind between the time horizons of newsmaking and the time horizons of global change. Journalistic time is not universal, and we have to decide which temporalities we wish to speak to and what the effects of those choices are.

Wicked problems and making time for journalism

The negotiation between everyday survival and far future sustainability, or between instant media gratification and long-term quality, are at their root 'super-wicked' problems (Lazarus 2009). Doing things which are ostensibly good or necessary can lead to bad long-term impacts, which breaks down established relationships of action and consequence. What we do with good intent can become our undoing, but we also have to deal with long and entangled networks of effects. The Anthropocene does things to time and the timescapes of modernity that leave us confused, which makes governance challenging as risks and outcomes are constantly weighed against one another. The Anthropocene is bigger than any of us – older than many of the solid institutions of contemporary society, and destined perhaps to outlive them – yet we all contribute to it piece by piece. We can always persuade ourselves that what is allowable today does not have a real impact on the future, so used are we to seeing the future as something that will happen to us rather than something we co-create on a daily basis.

Super-wickedness asks us to constrain the present in order to protect the future, or to live in two time periods simultaneously. Things which we believe to be to our benefit today can have hugely negative consequences in the future, and produce externalities which return to worsen conditions at some future point. At its most basic level this could manifest as a drive-through restaurant arriving in an area and creating jobs, but also leading to traffic congestion and air pollution that negatively impact on the local community. Even sustainability initiatives can have unseen or unintended consequences, so that when we report on them in an enthusiasm to encourage transition we can become blinded to their long-term consequences. Wind turbines can disrupt important carbon sinks, electric cars can leave toxic legacies and involve huge mining operations, and campaigning against urban development can have negative consequences if that demand and capital then manifests as suburban sprawl.

Journalism, then, is itself caught in a wicked problem, forced to mirror the society that creates it while bearing a moral responsibility to transform that self-same society in the near and far future. If we accept that journalism is modernity's sense-making practice, this lets us recast journalism as fundamentally about 'making time', about creating structures of sense-making that meaningfully communicate the interplay of deep and crisis time in an Anthropocene context. Constructing temporal frameworks in journalism is something that journalists implicitly already do, so what is needed is to change the temporal reference points and cognitive horizons of such work. As the ice begins to melt and we see the deep past and the recent past flow into view together, the severity of where we are right now and the complexity of the entanglements with the past and the future are self-evident.

8

Resilience and the Never-Ending Shock of the Present

The journalism industry in the 2020s is a strange mixture of resigned pessimism and radical technological promise. The broad trends in journalism over the past two decades – of atrophy, rationalisation and the chimeric spectre of equilibrium once the industry has come through a rough patch – show no sign of abating, while we turbocharge the causal factors and fail to take the steps necessary to emerge out the other side in one piece. At the same time we are confronted with declarations of innovation emerging from academia and the media industries alike, each with claims to transformative knowledge and potential solutions that become focal points for survival and continuation of the often precarious alliance between owners, consumers and journalists themselves. The parallels with climate change policy are obvious, and just as with climate policy, the need to survive as structures shift and disintegrate all around us has led media companies to embrace the idea of resilience in their quest to protect profitability and their centrality as key social institutions.

From its beginnings in the world of alternative technology and green social theory, resilience as a concept emerged into the mainstream in the 2000s as a tool for examining the sustainability of systems and practices to cope with the inherent instability of liquid conditions. This has been applied not only to institutions and infrastructures, but also to the global economy and the natural world. Since then, the resilience idea has been co-opted into a variety of new roles, from ersatz occupational health and mental wellbeing to managing stock market fluctuations, and as a euphemism for cost cutting and asset diversification when businesses feel the need to lay off staff. As the geographer Ben Anderson (2015: 60) surmises, resilience is now everywhere as a doctrine, an ethos, a programme, a buzzword, a normative or ideal concept, and even as a strategy for power. In a world of seemingly endless disruption and exponential change, the resilience narrative promises to arm us with the weapons to individually and collectively persist and survive when the ways of living and doing that we are accustomed to are pressured and destroyed.

Climate change has pushed discussions of resilience to the fore, as political leaders for their part attempt to define what is manageable and safe in terms of global temperature rises, and as climate scientists and policy makers working under them seek to calculate how vulnerable certain regional and planetary systems might be to total collapse. From the 2020s there has been a marked shift in policy discourse at the UN climate summits, from an emphasis on mitigation of climate change to adaptation, and a push for academia and governance to come up with exit strategies rather than preventative strategies. This is due in no small part to the broader embrace of resilience thinking across society, and in journalism too we can trace how resilience has become a guiding force as a management philosophy, handed down as a coping mechanism to survive in the liquid conditions of contemporary media work and the mental and physical attrition that involves. This manifestation – of resilience as a survival strategy pushing journalists to keep going in increasingly brutal conditions – is also found across academia and other public institutions, where doing 'more with less' or adapting to changing financial frameworks and environments has been internalised by management, and consequently also in many cases by staff.

Journalism itself engages with resilience as subject matter in its reporting of disasters and the consequent recovery and return to 'normality' which is promoted afterwards (Houston et al 2019). As a consequence, journalism has a crucial role in the mediation of disaster and in the propagation of resilient ways of thinking, an area in which it can go both right and wrong as it falls prey to the temptation of prematurely declaring crises to be over. Given the systemic shocks of climate change now making themselves felt, all journalism occurs in an unfolding global disaster zone to differing degrees, while internally also being a site of collapse and disaster itself.

The underlying question in both of these arenas is how real and realistic the resilience that is supposed to help us navigate systemic failure actually is, and whether resilience thinking is a solution or a trap. If it is to function as more than a management philosophy or a means of apologism for the systemic failings of the status quo then its worth is limited, but if resilience perspectives can help us to establish a path for journalistic survival then they are worth listening to. This duality is at the heart of discussions of what journalism might do, and how, in times of immense change, and how it can survive beyond the boom and bust of quick fixes and newsroom pivots.

Disruption and the illusion of sustainability

The hunt for resilient models has gone hand in hand with the emergence of a raft of news sites broadly grouped as 'digital native journalism'. These consist of outlets with no initial links to legacy media which have put their tanks on the lawn of traditional giants and made claims to be a new form of journalistic practice that is more fluid, reactive and innovative than its older stablemates. Typified by outlets such as Vice, Buzzfeed and the Huffington Post, in Britain, America and other markets where these empires initially expanded, these new digital mastheads were lauded for their approach to news from a digital first perspective, and have been identified as prime examples of digital news taking responsibility for climate coverage from legacy media (Painter et al 2016). This is due in no small part to their engagement with younger demographics and self-aware claims to be more in touch with the youth than their legacy rivals, but also a sense of wonder in a generation of media scholars still enamoured with the utopian potential of the internet and industry watchers keen to be aligned with emerging trends.

Many digital startups – often buoyed by investments from hedge funds and parent companies – pretended outwardly that they had found a business model to save journalism and push a future expansion of quality online content. Boasting radically different internal structures and mastery of digital income streams based on unprecedented numbers of readers and the potential capital they represented, these new entrants encapsulated a vision of journalism lighter and faster. As it turned out, however, many of these unbelievable success stories really were too good to be true. Just as people were falling over themselves to praise the dynamism of a new generation of young political and environmental journalists emerging on the platforms, the businesses themselves collectively embarked on a series of layoffs, responding to pressure as their funders demanded profit over quality, and attempting to cut their losses.

The rise and fall of Buzzfeed and its imitators is a cautionary tale about academic and industry co-option and the enchantment of digital capitalism as much as a lesson in journalistic sustainability, but Buzzfeed and its peers are of real historical importance in our understanding of media capitalism and how it determines what we see and how we see it. The relatively short life of a handful of sites in the Anglo-American world at the beginning of the 21st century exemplifies a hubristic moment in the datafication of both journalism and society, which is of extreme relevance to the larger ideological flows of capitalist modernity. The enthusiasm for and ultimate failure of these projects, we might argue, showcases how faith in disruption and newness obscures the need for structural reform and regulation, and indeed prevents sensible decisions being made that might prevent systemic collapse.

Each of these outlets had already undergone significant restructuring by the time Buzzfeed acquired the shell of former rival HuffPo in 2021 and announced yet more redundancies in a further cannibalisation of the sector. The fast growth of the digital outlets and their engagement with audiences began to emerge as a speeded-up iteration of the path followed by legacy media, with familiar sacrifices for the sake of profit over quality or sustainability. Obsession with the future of news as a form of corporate planning, not to mention other forms of quality journalism beyond news reporting, can be seen to exaggerate this effect as owners and shareholders seek to rationalise and safeguard the profitability or viability of their assets. Yet even more community-driven projects, with less financially self-interested owners, have found it difficult to maintain financial momentum in a transient digital market.

The Correspondent, a membership-driven startup which grew out of the Dutch-language *De Correspondent*, embodied the same spirit of newness and disruption, raising over $2 million in seed funding in 2019 prior to its launch. The Dutch parent described itself as a laboratory for journalistic renewal, but its anglophone experiment was to be wound up again two years after it was born due to declining reader revenues and mounting liabilities. Having been celebrated as an alternative to the mainstream digital model of value extraction and click-chasing, the failure of *The Correspondent* and its self-styled global community provided a sobering reminder that good branding and a tidal swell of optimism are not in themselves a guarantee of resilient practice, and that disruption must at some point transform into institutional stability.

The rise and fall of both corporate digital natives and nonprofit media relying on considerable emotional labour exposes the difficulty of establishing resilience within the frameworks of online commerce and interaction offered up by the internet as it currently exists. Among the failures there are some successes, but what is clear from even a brief glance at the digital native news bubble is that technological innovation which promises revenue and sustainability through its inherent difference and entrepreneurial allure is still subject to the same constraints of political economy as legacy media, whatever its outward politics may be. But despite the documented problems of digital journalism projects, the past decade has also seen the emergence of new networks of environment- and climate-specific outlets, and in some cases individual journalists working alone with annual budgets equivalent to a reasonable professional wage thanks to sponsorship platforms and loyal readerships. There have also been significant investments in climate journalism from major media companies with deep pockets such as *Le Monde*, the *Financial Times* and the *New York Times*, where institutional strength facilitates well-resourced spin-outs.

Climate activists have also become media personalities and content producers in their own right, coordinated through formal and informal networks which become particularly visible around events such as the UN COP summits and election cycles. One of the more notable features of this has been a convergence of press work by established globally active NGOs with backing for significant investigative journalism and fact-based public service journalism.

Greenpeace, who since their founding in the 1970s have come to play a major role not just in environmental protest but also as a source of information and education, have added news production to advocacy through Unearthed, an investigative journalism platform publishing text and video content highlighting environmental abuses worldwide. Grown from an initial project known as Energydesk, the operation was able to mature into a significant quality provider of climate news supported by international financial clout and hundreds of millions of dollars of capital.

Like multinational media businesses, NGOs are able to take advantage of the footloose and globally expansive nature of their work and choose where to base themselves based on prevailing conditions. Whereas international media organisations may base this decision on the relationships they have with governments, the state of antitrust legislation and tax levels in specific places, new types of transnational institution are able to operate from places where conditions suit them and – ironically – take advantage of the international flows of capital and organisation that underpin many of the businesses they seek to expose and monitor.

Greenpeace's engagement with climate journalism illustrates well the longer term investment climate and environmental coverage possible without an immediate need to satisfy shareholders, or indeed to match income from journalism to the outlays of doing journalism. This may bring its own ethical problems in terms of accountability and editorial independence regarding Greenpeace itself, but Unearthed was able to achieve major victories in its first few years.

In a notable coup, the platform unveiled film of Exxon lobbyists openly admitting to opposing emissions reductions and climate measures in the US Senate and Congress, something made possible not just by their journalistic presence but by the financial and legal muscle they enjoyed which allowed them to do so safely without fear of legal reprisal. Starting new media platforms is extremely difficult, and finding an audience for them even harder. The hedge funds and angel investors of new media may have both wealth and power, but these behemoths are not institutions in themselves; rather they have an almost pathological need to undermine and subvert institutions for the sake of their own dominance. Resilience is less a question of riding the waves of change than building strong foundations from which to build and rebuild.

Public goods in the futures market

In the dissonance between the vultures of venture capital picking over the remains of failed projects and desire by journalists to produce good work, there emerge alternative visions of the media future that help us to imagine new institutions.

It has been suggested that in the future good journalism will be a question not of large capitalist media structures and the zombie institutions of old media, but of values embodied by journalists themselves operating in spaces beyond traditional media modes (Deuze 2019). Similarly, there been much discussion of 'pop-up' news culture inspired by the economic organisation of the maker movement, running specific journalism projects for specific time periods or events, with quality and breaking even the main priorities (Wall 2015, 2017). Such practice envisages journalism as a series of time-limited interventions in the fluid public space, mimicking and embracing the transience of liquid working practices and the wider project-based thinking of the business world. This is a scenario in which the continuous practice of journalism is divided into a series of discrete projects which can be crowdfunded or run using grants for a set period of time without the overheads of legacy media or demands of shareholders, and in which temporary utopias come into being before dissipating.

In a world of seemingly perpetual upheaval and collapse, this embrace of the short half-life of contemporary journalism might appear an appropriate response to (con)temporary conditions, a place where the fast-moving ethic of contemporary tech capitalism subsumes the DIY instincts of radical outsider media to create a hybrid craft that is subject to constant evolution and reinvention.

This is an enticing vision, but if we see this in terms of a broader polycrisis of institutions of modernity common to the Anthropocene we can also look beyond this narrative of adaptation and hacking the present. It leads us to realise that resilience as an adaptive strategy to the natural laws of the market is in fact a sleight of hand by those who wish to see the status quo broadly unchanged. We find familiar questions about the political economy of media and classical concerns about the organisation of labour, often masked by an enthusiasm for innovation and newness. The claims of short interventions to be a solution to structural problems in fact normalise the transience of the media space, subscribing to the same narrative of flexibility and speed favoured by asset-stripping conglomerates, and the maxim of exponential change, lighter, quicker and faster.

Recognising the multiple threats to the future of information provision, John Hartley argued for journalism not just as a convenient feature of liberal democracies but as a human right and social utility which required institutional reform (Hartley 2008), by embedding journalism in core civic institutions rather than as a quasi-autonomous institution of its own. Hartley returns to the principles embedded in the 1948 UN Universal Declaration of Human Rights and its emphasis on the ability to 'seek, receive and impart information' as a precondition to human freedoms.

Hartley's argument is that human rights notionally apply to all people, everywhere, all of the time, regardless of age, race, gender or nationality, and that this can also form the basis of universalist media. The idea of human rights as it emerged throughout the 20th century has been used to fight big business, compel governments to give shelter to refugees, and to bring climate change litigation to court. In a time of global need, even a loose global legal regime is an asset, especially when backed by powerful transnational institutions. Why, so this line of thinking goes, should we pursue journalism any less than human rights or international law? Even from a realist perspective, this can help us to see a media which – to paraphrase Dag Hammarskjöld – does not bring us to heaven but at least saves us from hell.

Reversing the emphasis on adapting to market conditions places journalism at the service of wider aspirations for global society. Just as the United Nations has become an (imperfect) framework for discussion of and action on climate change and biodiversity, returning to these core principles of globality opens up the possibilities of global interventions in the future of journalism around the fundamentals of representation and participation. At its root, this is a question of civic structures and ownership of technology and the public space. It feeds into discussions about the nature of digital capitalism and information provision itself, but it also means engaging in open conflict with the power holders who benefit from the private holding and financialisation of public utility.

Like the oil industry, technology companies lobby governments hard to prevent changes which would be detrimental to their business models, as happened on both sides in 2021 when Australian news sites were prevented from posting to Facebook in protest at legislation designed to ensure payment by social media giants for the news content they use. Such conflicts lend themselves to David vs Goliath narratives of popular resistance, but the main beneficiaries of the Australian stand-off were private media companies, who were able to negotiate payment deals with platforms in a continuation of duopolistic cartelisation. Given the enduring issues in the economics and management of journalism worldwide, many attempts to save journalism can be seen merely to co-opt competing forms of profit-seeking media hegemony, with little discussion of the wider political economy of media or the needs of the public. Survival for media organisations does not necessarily entail survival for expansive journalism, running the risk of preserving and replicating negative existing structures of media ownership and the cartelisation of information.

Resilience for shareholders or thought leaders does not mean resilience for journalism itself. The future of journalism as both a distinct practice and method of mass communication, then, depends on the creation of strong institutions and norms, and ensuring its survival as sense-making practice in the flow of modernity relies on assembling coalitions of interest, cooperation and transnational regulation in the same way that we are forced to respond to climate change. Journalistic governance embodies many of the same dilemmas about the conflict between a liberal ethic of competition and market supremacy and the overriding and apparent need for coordination, mitigation and strong institutional leadership in the common interest.

As Catherine Happer (2017) has argued, when media begins to internalise the logic of financialisation, and by extension its versions of the future, it not only reports according to the values and norms of a more broadly financialised society, but locks in a reliance on specific conceptualisations of how the future might be and what is perceived to be normal. Indeed, taking the 2008 financial crisis as an example, Happer traces the way in which recovery and a return to normality were portrayed as a question of replicating conditions prior to the crash and reducing the events of 2008 to bad governance by a select group of individuals. Here there are clear avenues into how we discuss both climate change and journalism, suggesting how potential solutions to both are severely constrained by the adherence to existing institutional settlements.

All of this helps us to interrogate the dominant narratives which emerge about the future of journalism. The first is of brave innovators, using models such as single-person campaigns relying on crowdfunding and newsletter formats, bypassing mainstream media. This ideal of an open commons of individuals operating in constellation is an attractive one in many ways, embodying as it does an ethic of homespun earnestness and anti-establishment subversion.

The second of these is the idea that business as usual can somehow be regained, with news organisations reasserting their place as the fourth estate in the international order of liberal societies, safeguarded as an integral part of a free thinking and economically free global condition. Under this scenario, intervention and regulation help to stem the decline of media as a profitable enterprise, allowing news outlets to fulfil their role as watchdog and commentator.

The third scenario is one of exponential change, in which technology and innovation on a wider level than individual entrepreneurship is able to come up with a solution to the enduring problems journalism faces. In an obvious parallel with climate change solutions, these seem to rest on the use of technologies which either do not currently exist or have not been tested at scale, a world of perpetual beta testing and cautious optimism which present themselves as the future of news.

It is instructive here to look at how subscription-based online streaming services have 'saved' the music industry from illegal downloads, but are rarely in themselves profitable (Hesmondhalgh 2021: 4), requiring funding from venture capital or parent companies and often offering meagre returns for producers. The manner in which streaming is publicly seen to have solved issues of viability despite stories of destitute artistic labour and corporate data harvesting encourages scepticism about the potential for journalistic fixes along the same lines. In the case of both music and journalism, consumers do not always realise the financial mechanisms or motivations behind what they consume, not least the complex processes of hedge fund investment and financialisation behind services they think they are paying for.

In terms of what is likely and achievable in developing resilient journalisms, it is not unreasonable to envisage a synthesis of the perspectives outlined earlier, but in the context of broad institutional reform. Each of the three scenarios described here is an articulation of the contemporary liberal economic settlement, but what each of them lacks is the demand for institutional transformation which might ensure truly resilient practices and allow journalism to survive with its dearly held ideals intact. This institutional strength is intimately bound up with the notion of community, represented on the one hand by more traditional ideas of a community of consumers, and on the other by ideas of a post-material or post-consumer journalism in which readers engage with platforms beyond immediately transactional ties.

A return to strong civic institutions is a cornerstone of post-neoliberal thinking in which media is reclaimed from conglomerate ownership, but the utopian ideals of the public interest startup are reliant on a specific conception of agency in the public space which is also built on European and North America norms.

Media and the new institutionalism

Fundamentally, effective institutions are critical to collective action and journalism is in need of global institutional support in the same way that climate change (rightly) provokes an international response. Casting around for answers to these problems, the political scientist John Dryzek (2016: 955) concludes that 'taking the Anthropocene seriously suggests an evolving institutionalism joining inquiry and practice, in the face of existing dominant institutions that fall so far short of the requirements of this emerging epoch'. Taking the future of journalism as a mode of interrogation seriously requires engagement with the shortcomings of existing institutions and the engineering of new ones which can respond in times of challenge.

We can also see this need for new institutions of media in terms of a broader move towards neo-institutionalism worldwide and reappraisals of the market forces of the past 40 years. There is an increased recognition that developmental and civic challenges have to be pushed and maintained through strong institutional organisation, with a consequent resurgence in forms of social democratic economics and interventionist politics in the work of economists such as Mariana Mazzucato (2011) and her recontextualisation of global economy in terms of strong public investment and regulation, or of Brett Christophers' (2023) critique of the asset management industry as a shadow-class of global control. Mazzucato and other economists who rose to prominence after the 2008 global financial crisis have sought to re-evaluate the importance of public and mutual institutions in the creation of the future, arguing that strong public (if not state) institutions are necessary in order to maintain viable commercial ones, but also that the strength of such institutions actually facilitates risk taking and slow development that the private sector is not capable of.

Ultimately, the question of resilience and survival for media in a turbulent world is teleological, rather than just a question of adaptation and accommodation or market solutionism. Truisms about a changing planet, a fast-moving global situation, about visionary leaders and the shock of the new may speak to the managerial desire for constant proof of dynamism, but this is not necessarily helpful in the current predicament. Instead of seeing media atrophy as a natural process to which we must adapt through flexibility and innovation, such changes can also be viewed within the wider potential for reappraisal of the functions of political and civic institutions in a post-neoliberal landscape. Because the Anthropocene problematic invites us to think big about the broader political economy of the world we live in, it also asks us to reconsider the informational structures we inhabit and how they are financed and constructed. Only through doing so, perhaps, can we create genuinely resilient journalisms.

9

Media Cosmopolitanism in the Anthropocene

Journalism and journalists like to imagine themselves as archetypal cosmopolitans, able to bring together disparate and distant places as constant characters in the Story. Through a combined ethic of objective distance and insider curation, journalism's ability to enlarge individual worlds and shrink the globe have a potential to speak to many of the key concerns of the Anthropocene, a task at which it sometimes succeeds but often fails miserably. On a warming and interconnected earth, cosmopolitanism is more important than ever before, causing us to ask necessarily sceptical questions about journalism's existing claims to globality and omniscience as a solution to the problems we face.

In the past few decades there has been a growth in university programmes seeking to produce global journalists capable of operating within and across the transnational media landscape, based on the understanding that the economic cosmopolitanism of globalisation requires enthusiastic workers in the global informational marketplace. This functional cosmopolitanism, which is also found in the most shallow manifestations of European integration and from the UN climate COP summits to the British Airways business lounge, is often shorn of its inconvenient conflicts and contestations, and its vision of the international and the transnational can be decidedly unimaginative. Indeed, in the manufacturing of a global cosmopolitan class, it is easy to feel that we are all in touch with each other when that class is by design self-limiting and can be more akin to what Craig Calhoun (2008) has called 'an attitude, a style, a personal commitment' than a meaningful engagement with the planet.

Understandings of cosmopolitanism as they manifest in public life and sociology are historically and culturally rooted in the way modernity has constructed the globe and the cultural interactions between its disparate parts, complete with relationships of power and subjugation. The cosmopolitan thinking which emerged in 19th-century Europe as technologies of communication and travel brought the continent closer together morphed into another brand of European cosmopolitanism after the Second World War, with the rhetoric of European entanglement and interdependence as a political project and failsafe mechanism against a repeat of the horrors of the continental and global conflict. The European Economic Community and European Union (as it later became) were reliant on the intellectual force of European cosmopolitanism as an ideal, from both left wing and more liberal perspectives alike, and even today the memory of the Second World War and the Cold War are foundational to the European mode of cosmopolitanism, expanding the world and shutting it out at the same time.

In Germany in particular, Jürgen Habermas emerged as one of the leading proponents of cosmopolitan thinking, building on the work of Immanuel Kant, and this legacy is still strong in German public life and the corridors of European power. In the European project the sociology of cosmopolitanism and postwar rapprochement has provided intellectual cover for the European Union's utopian and moral claims despite ample evidence to the contrary in areas from refugee policy to global trade justice, debt forgiveness, overseas military campaigns and banking collapses.

This tradition of European cosmopolitanism and solidarity was both developed and problematised as part of the sociology of modernity by Ulrich Beck, whose cosmopolitan theory looks to deal more clearly with forces of globalisation in the 21st century while staying true to the core ideals of a social liberal European project (Beck 2006). Beck suggested that under the conditions of globalisation cosmopolitanism cannot merely be willed into existence, but requires consistent political interventions to cultivate a cosmopolitan perspective and to tackle inequalities and injustices while building social solidarity. This is true even in Europe, where the rhetoric of harmony and harmonisation emanating from the European Commission collides with a crunch with member states, lobbyists and the European Central Bank.

In the Anthropocene, however, cosmopolitanism is more problematic than its utopian ideals and proponents might have us believe. This is not least because cosmopolitanism as broadly understood is often the privilege of the few, and the overlapping but fundamentally differing concepts of cosmopolitanism, internationalism and globalisation are often lumped together for the convenience of those making the rules. Because the ideal of the footloose cosmopolitan is bound up with specific ideals of mobility, access and perhaps most crucially detachment, we cannot merely ask for more cosmopolitan thinking as an answer to Anthropocene problems. A news reporter airlifted into a flooded village in Bengal might interview a local with a fraction of their economic and cultural capital who cannot live the same cosmopolitan life as the reporter, and indeed where the globalised presence of the reporter is of little benefit to the person whose village is underwater. Claims to globality are often merely hollow representations of other people's tragedy, with the powerful and the mobile dressing their exceptional status up as universality.

Nowhere is this seen better than in air travel, the cosmopolitan mode of movement par excellence, in which 50 per cent of aviation emissions are produced by just 1 per cent of the global population (Gössling and Humpe 2020). Mobility even for some, let alone for all, is an incredibly resource-intensive undertaking. As Mimi Sheller and John Urry (2006) outlined in the idea of the new mobilities paradigm, energy and mobility are intimately linked, and those of us who are most mobile are also the biggest consumers of energy.

Cosmopolitan theory has evolved in line with globalisation and its discontents, with sociologists such as Gerald Delanty (2009) and Gurminder K. Bhambra (2016) pointing to both the eurocentrism of cosmopolitan theory to date, and the need to recognise parallel and different types of cosmopolitan societies beyond the traditional ideas, however well intentioned, of Euro-American cosmopolitan thinking. Simply to be globally minded is not enough, nor is there a neat division between liberal cosmopolitanism and parochialism, but rather frictions and alliances between different conceptions of how the world is and should be.

Elites and cosmopolitanism's enemies

In an age of climate conflict, where the values of openness and global understanding jar with political cultures of isolationism and denialism, we also see, in Beck's (2002) words, the enemies of cosmopolitanism. These enemies take the form of nationalism, neoliberal globalism and democratic authoritarianism, with its impulses to control evident in the way governments in the Global North have taken to heavily policing both migrants and their own populace.

In the contemporary political space cosmopolitanism is also an insult as much as a badge of honour in certain quarters. Journalists can stand accused of being too cosmopolitan by their detractors, and stereotypes of the liberal New Yorker, the North London media worker or the Sao Paolo TV professional reinforce the idea of journalists as a separate class, giving ammunition to populists and isolationists who weaponise discomfort with globalisation and disembedding of communities to blame metropolitan elites. In 2016 the then British Prime Minister Theresa May caused uproar when she asserted that 'if you believe you are a citizen of the world, you are a citizen of nowhere' (May 2016), taking aim at both the cosmopolitanism of the European project and those who would deny the situated identity of Britishness (or rather Englishness), and by implication the legitimacy of British nationalism and patriotism.

Cosmopolitanism as an ideal now finds itself at the centre of the global culture wars, a threat to some and a cause célèbre for others. Beck (2002: 18), following on from Kant, theorises cosmopolitanism – or rather the process of *cosmopolitisation* – as processes of both internal and external inclusion which can cross boundaries as well as put them up. According to Beck, external cosmopolitisation includes coming to terms with the otherness of nature, the otherness of other civilisations and modernities (including, from a Euro-American perspective, African, Asian and South American modernities) and, most crucially perhaps in terms of Anthropocene processes, the otherness of the future itself.

The cosmopolitan gaze is one that is capable of seeing these others and making them part of the everyday. To some extent this happens automatically because of globalisation – what Beck and others refer to as 'banal' cosmopolitanism whereby we are aware in our everyday lives of spaces distant from our immediate surroundings. This is typified in the 'cosmopolitanism' of eating sushi, listening to K-pop, watching a YouTube ASMR video from Russia and streaming American films in the space of a few hours. This is cosmopolitanism stripped of its high cultural pretensions, that goes hand in hand with the general convergence of media and the overlapping plates of mainstreams and subcultures. At once democratic but equally chaotic and directionless, a teenager watching Russian ASMR videos is as cosmopolitan as a fund manager on a flight from New York to Paris watching Bloomberg to keep up with the latest market news.

This tension between banal and elite cosmopolitanism is not binary either. The elite cosmopolitan will also engage in acts of banal cosmopolitanism, and those of us who more usually engage in banal acts of cosmopolitanism may occasionally stray into the practices of the elite. We can also successfully ape elite cosmopolitan behaviour, persuading ourselves that we are participating in globality through the performance of worldliness and the consumption of global experiences without meaningfully engaging with the wider world.

This also invites consideration of what cosmopolitan journalism actually is, and how we pin it down. The journalism scholar Sarah Van Leuven has described how global journalism exists somewhere between dream and reality (Van Leuven and Berglez 2016) as a dearly held ideal which falls down in the face of national news agendas and competing interests, and we can say the same of cosmopolitanism itself, a utopian ideal in the minds of editors and foreign reporters that often ends up falling back on the narrow frameworks of its audience and their expectations about the world at large.

Paternalism and the global citizen

We can also question whether this cosmopolitanism is based on representation or projection. Certain schools of global journalism participate in a form of cosmopolitan paternalism in which journalists, as archetypal cosmopolitans, underpin an elite cosmopolitan project. This idea of the world as a steerable system with the cosmopolitans in charge – what the anthropologist Ulf Hannerz (2004: 205) describes as a process in which 'strategists, who are responsible for shaping the world, and journalists, who are in charge of explaining it, must learn globalist thinking' – feeds into the dream worlds of liberal globalisation, and of the informed (rather than the rich) as the holders of true power. This approach speaks well to the world of think tanks, policy forums and leadership conferences which characterise contemporary international politics, spun out into podcasts, editorials and breakfast workshops. It is a world in which responsibility rests on an elite capable of making informed choices with the best interests of a loosely conceptualised global community at its heart. This dominant form of transnational thinking is a place in which issues of class, equity and politics are put to one side and replaced with a hunt for 'solutions' and strategies, and where globalisation is portrayed as an organic process which should be curated rather than challenged.

In this paradigm, journalists are often invited to participate in this ersatz cosmopolitanism as communicators, facilitators and talking heads. They are seated at the top table at the World Economic Forum, behind the scenes at institutions of global governance as panel participants and moderators, and as commentators on the types of reforms and actions such leaders ought to take or the situation on the ground in the pages of *The Economist* or *Foreign Policy*. In this globalised conversation, some journalists are styled and style themselves as public intellectuals who can give an informed opinion on the situation in Russia at one moment, and talk about China and climate change the next, in a carousel of analysis.

Tellingly, the *New York Times* hub at the Glasgow climate negotiations – a lavish warehouse conversion filled with delegates wearing lanyards – invited the public to draw on the expertise of *New York Times* journalists alongside a who's who of international policy and human rights voices who operate in what participants refer to as the 'climate space'. When journalism has shown its limitations, some may even make the jump from hammering out copy to working as special advisers in government, or in multinational companies where their nominal insight and ability to convince is a sought after commodity. When in 2015 *The Guardian* announced an anti-oil campaign called 'Keep it in the Ground' under the direction of then editor Alan Rusbridger, it did so with conviction about elite cosmopolitan journalism's moral and discursive force, actively embracing but also overstating its position as a major driver of change, built in part on its self-image as an important voice in British public life, and through this Britishness as an important global voice.

For news brands with readerships who see themselves as globally aware and politically engaged, a powerful cosmopolitan brand remains critical to market positioning, as is the ultimate evolution of the newspaper in question from a left wing British regional to a digital news service for a slice of the progressive Global North. Endorsement for such globally minded news brands functions as a form of specific cosmopolitan expression without fundamentally disrupting the underlying organisation of British or European cosmopolitan thinking, and crucially without delving into the complexities of the global fossil economy as anything other than a morally compromised industry.

Global participation and media labour

Cosmopolitanism also looks different at the centres and peripheries of media work. Behind the handful of big name international correspondents, foreign journalists are quite likely to be freelance, working on a story-by-story basis and living unglamorous lives, also doing second jobs from language teaching to public relations and corporate copy generation to pay for their apartments and childcare costs.

Though there remains a select group of well-funded international correspondents on the payrolls of major global print and digital brands and in broadcasting, this most visible component of foreign journalism is neither typical nor especially well suited to its task of cosmopolitan entanglement. The dream of the *fin de siècle* Viennese coffee house or the well-stocked hotel reading room still overshadows our notion of what international journalists should be, but in a world of fast news we must look not to people putting together meticulously prepared dispatches, but to a global media workforce caught in the same liquid flows of globalisation and modernity as everyone else.

In the idealised world of the cosmopolitan journalist we are all engaged in a single conversation, in touch with the forces that shape our world and with each other. Yet with the decline of well-resourced foreign journalism the ways in which the world manifests change too. What we have seen in the past decade or so, with the phenomenal decline in foreign desk and editorial resources, is the rise of the journalist as individual cosmopolitan actor rather than as a technician in a regimented vertical network.

This is the case for the freelancers in particular, who are competing in a buyer's market for commissions, but this all creates a narrative problem. By conforming to a loose market principle of the journalist having to win work, the costs of their training are externalised and potential reporters are expected to arrive with their product fully developed. As part of a general trend towards placing more of the costs of work on journalists rather than outlets themselves, the skills needed to perform these liquid functions of the global journalist become the preserve of people with time and resources, embedding inequalities in the journalistic ecosystem.

As part of the general participatory condition of contemporary media, what was once seen as a form of radical practice has been normalised. Diverse publics are brought in and contribute through both paid and unpaid labour, but journalists themselves also end up living in a twilight space between the professional and the merely participatory.

Citizen journalism has its roots partly in the emancipatory can-do ethic of tech startups and DIY media, but also carries an altruistic belief in the democratic value of 'normal' people having their voices heard without the gatekeepers of traditional media. Sure enough, anyone can set themselves up as a journalist today and perform that role in the public space, with the social media handles of young journalists seeking to solidify their aspiration before a single penny has entered their bank account. The visible online journalist engages in incessant rounds of pitching until something sticks, but also through blogging and prominence on social platforms as an expert in a certain field and an audition for the 'real' work to come.

The result is a constant field of semi-professionalised journalistic labour which blurs the lines between reporter, professional, hanger-on and much else. The prevalence of free labour in journalism has drawn the attention of researchers, who have codified free work in terms of defined roles: The Slave, The Carer, The Apprentice, The Prospector, The Hobbyist, The Volunteer and The Patsy (Fast et al 2016), seeking to build a picture of creative labour which is at once highly visible yet entirely unrewarded. A carer is a media worker who may work for reasons of altruism or community benefit, and exercises such as climate blogging or community information provision can function as forms of care, while a patsy is someone who unwittingly performs unpaid work that generates profit for someone else. The foremost example of this is probably the tendency for online and broadcast media to aggressively hunt for eyewitness video and images on social media and then to engineer its use without paying, in effect using members of the public as free news gatherers. This also extends to professional journalists who may happen to be in the area at the time or have a particular expertise.[1]

These distinctions are important because they expose the power relationships in the nominally open field of participatory journalism and the idea that all of us can make the news. The dream of a global journalistic commons in which we are all cosmopolitans, in touch with one another, meets the reality of the political economy of globalised content production.

Cosmopolitanism stumbles when equity is on the table. For media cosmopolitanism to be meaningful, it must not only seek to bring in perspectives from elsewhere but allow those perspectives to contribute to the generation of reporting itself. In 2019 the *New York Times* attracted criticism for a job advert for its Nairobi bureau chief offering adventure and intrigue, including terrorism, the 'pirate seas' of the Horn of Africa and the forests of the Congo. The subtext was that Africans need not apply even though

Nairobi is a centre for African media, and that the stories the bureau chief should generate were to fit a particularly American idea of cosmopolitan foreign journalism in which the employer's money would flow not to a local journalist but an enterprising expat.

What we see when we begin to break down journalistic labour is that the structures of global interaction replicate the organisation of labour and work, replete with implicit and explicit nods to class, education, citizenship and ethnicity. This is reinforced and perpetuated by control of modes of distribution that are themselves bound by norms, as well as by audiences. Elite news services are inevitably bound by their own real and imagined readership, so that even a globally minded audience must access the world through the prism of people like themselves.

Occasionally there are upstarts and mould-breakers who do foreign and global media differently, but rarely do consumers find their way to these outlets for more than the odd article, and even more rarely do these outlets and the people behind them have the ability to reach audiences who might be interested in what they have created. Even more tellingly, in areas that have gone 'dark' due to a lack of resources from international media organisations, hobbyists and semi-professional expat media have picked up the slack.

Fundamentally, grasping media work is critical to understanding how and why things appear in the global media space, and it is crucial to deconstructing how cosmopolitanism as an ideal manifests. We may all be participating at some level, but how real that participation is remains questionable. When we put journalism into the public space in the hope that someone might read it, we are more often than not engaging in rituals of volunteerism and hobbyism rather than work itself, and we find ourselves expendable to those at the centre of media power.

Climate change and the new cosmopolitanism

Old cosmopolitan modes are not fit for purpose, and one of the effects of Anthropocene events and Anthropocene thinking is that we are by necessity forced to confront the translocalism of the things we see and do, but also to confront and understand the increasing foreignness of the future itself and the transformation of our own localities into less recognisable forms.

Beyond its attempts to solicit commitments from different governments, the IPCC is less interested in nation states than in the behaviour of the environment across 26 regions of the globe which are broadly interdependent. These 26 regions both bring nation states together and separate them, so that while northern Europe is one region containing England, Scotland, Wales and the island of Ireland alongside Denmark, Norway, Sweden and Finland, the USA and Brazil are subdivided into three regions apiece, some of which are shared with neighbouring countries. The arbitrary nature of the nation state on the maps of global climate is exposed through such scientific framings, as climate effects are shared by multiple nations but also wildly different impacts can occur in the same political unit.

The US could experience severe drought in the Midwest but torrential rain, flooding and storms in the north-east. It is impossible to talk about 'the climate in America' in meaningful terms beyond saying that each of the parts of the US will experience different extremes. Equally, there exist relationships across boundaries. When wildfires in the US and Canada as a result of the Pacific heat dome of 2021 pushed particulates into the atmosphere, they drifted across the North Atlantic and led to spectacular red sunsets in Scotland and Norway, a visual cue that the horrors of burning towns in British Columbia were all part of the same system. The wilful isolationism over events in supposedly far away countries is no longer tenable.

These two little corners of the rich world, however, are nothing compared to the places where the majority of the global population reside, live and work. In the European imagination the spectre of Africans and Middle Eastern migrants moving north has been a stock bogeyman of immigration politics since well before climate change became a leading driver of the movement of people. Indeed the environmental movement in Europe has flirted with the interventionist idea of population control in emerging economies since the 1960s, and more conservative environmental figures have openly endorsed programmes that would reduce the populations of countries whose citizens might seek to breach the borders of Europe's internal cosmopolitanism.

Under climate change it is not just the prospect of economic improvement or political censure that drives people to migrate, but the sheer unliveability of circumstance. The International Federation of Red Cross and Red Crescent Societies (IFRC) estimated that 200 million people a year could require humanitarian assistance by 2050 due to climate-related impacts (IFRC 2021), with large numbers of those people being temporarily or permanently displaced.

Stories of refugees arriving on the shores of European countries in makeshift boats or crossing the US border by land are headline fodder, but their reduction to problems for northern countries to shoulder, however sympathetically, does little to reflect the situations from which they have come. Such narrative standpoints both diminish the experience of people living in such places and give a somewhat false sense of security to the solidity and unchangeability of Europe and the US. In truth, climatic change has implications for every single country on earth, and though this may result in migration northward, the effects of climate change are not exclusively to be felt in the countries producing these migrants.

The future is also going to be a foreign country without many of us moving an inch. Coastlines will change, forests could turn to grassland in the space of a generation and the standard of living may well fall in advanced economies as climate shocks begin to make an impact on the economy and the global food system. Even domestic journalism in the future will involve coping with the inherent difference of climate change as previous certainties become fluid; the procedural news of the domestic agenda will be contingent on world-spanning events of climatic disruption. The things which will happen on home soil will appear increasingly alien, and the formerly alien and foreign worlds of humanitarian crisis and emergency may become depressingly familiar.

Whereas previously foreign correspondence has relied on tools and arguments to engineer proximity, right down to the reductive question of whether any Americans or Britons have been involved or witness to events as per the wont of certain newspapers, this is transformed in an Anthropocene context. In the Anthropocene proximity and agency is implicit, as in serious floods in Germany in 2021 followed by the severe drought of 2022 where parts of the Rhine turned to gravel. Though questions were asked about the performance of the German government on decarbonisation and its readiness to deal with climate threats, the story was one of international necessity weeks ahead of the release of the latest IPCC report. Whether we like it or not, the world is on our doorstep.

Methods of journalism and methodological nationalism

Cosmopolitan journalism must also confront the failure of what in social science has been labelled 'methodological nationalism' (see, for example, Wimmer and Glick Schiller 2002; Beck 2006). This entails accepting that ways of doing and seeing things on a global level are often tied to the concept of the nation state and its behaviour in ways that are fundamentally unhelpful. In the same way that sociology has been criticised for too often focusing on the nation state as the natural social unit of modernity, in journalism the way stories are reported and the terms of their engagement are often rooted in national or regional traditions and not in an overall view of global conditions.

Claims to cosmopolitanism are limited by the language and ideals of the international order (and particularly the dominance of a few countries) as the primary means of understanding global processes. Phrases such as the 'Chinese Century' or the discourse around the UK leaving the European Union create false impressions of the independence of national actors in understanding the impacts of capitalist modernity and the global economy, while a model of international relations that articulates climate negotiations in terms of the self-interest of national parties obscures the fact that both the environment and the global economy have fault lines which do not correspond to international borders. Migration, climate change, food security and the behemoth of the interdependent global financial system are all transnational systems. Methodological nationalism is, as the journalism scholar Simon Cottle (2011: 310) describes it, a problem because 'Today's global threats go to the core of contemporary arguments about global cosmopolitanism.'

The problems of methodological nationalism for journalism mean that approaches to both practice and narrative often bottom out in terms of national media cultures, and implicit assumptions about the behaviour of audiences in particular too. This is best demonstrated in the dissonance in transatlantic journalism culture, whereby the American media and the British media diverge markedly in their approach to the coverage of topics based on enduring differences in the practice of press work itself, but conversely both also belong to an anglophone intellectual and media culture that shows commonalities when compared to the rest of the globe.

Here we can also find a false cosmopolitanism that claims to have global resonance and interaction at its heart. Norms engineered in Europe and North America are internationalised and thus rendered in the international space as 'global' ways of working. Much of the North American journalism community in particular assumes the US as the centre of excellence in journalism standards, and there are a multitude of programmes for journalists from the emerging economies or countries with 'new' democratic cultures to come to Britain, America and the EU. These schemes very often present the press in advanced liberal democracies as a transferable universal values system, reinforcing the greater dominance of the Global North in the production of globally accepted forms of knowledge. Such problems in the organisation of global journalistic knowledge are rarely challenged, but once seen they are everywhere.

Understanding journalism's own methodological nationalism is important to grasping some of the failures of journalism to date to grapple with global issues, and the way in which our own positionality as journalists changes the news we produce, but equally how we are forced to work within arbitrary frames of reference in the absence of any real singular international or transnational media sphere.

Towards an Anthropocene cosmopolitanism?

Given the failures of the shallow cosmopolitanism of globalisation, what then does Anthropocene cosmopolitanism look like for the media? The anthropologist Anna Tsing has spoken of a world in which the cosmopolitan gaze moves beyond the usual suspects to take in the full spectrum of life on earth, from the oligarchs of international business to civil society and indigenous groups, and on to the nonhuman actors of natural systems. This might seem fanciful, a nod to environmentalism's recurrent desire to lend the earth personality and agency in line with its own cosmologies, but from a systems perspective it also makes perfect sense if we see the natural world, humans, capital and politics as intertwined and interacting. As Tsing puts it:

> Imagine ethnic groups, corporations, refugees, nongovernmental organizations (NGOs), nation-states, consumers, social movements, media moguls, trade organizations, social scientists, international lawyers, and bankers, all swarming alongside creeks and earthworms to compose the landscape, to define its elements, carve its channels of flow, and establish its units of historical agency. (Tsing 2000: 327)

What Tsing touches on is the emptiness of a cosmopolitanism that we perform without paying attention to the full range of processes going on which shape this global modernity, noting that when you 'click on worldmaking interconnections, your world fills with global flows' (Tsing 2000: 327): an instantaneous engagement with the web of world-making interactions. Each interaction begs questions – how much, where from, by whom and why?

Stories of African villages under threat, of displaced indigenous peoples and conflict in the Middle East take place as projections of European crisis, but these places also have vibrant independent media cultures and potential futures beyond disaster and as a problem for the conscience of richer nations. The coexistence of elite and banal cosmopolitanism in journalism means that we are never far from globally important events, but making these visible is tougher. Cosmopolitan thinking can take the spectacular and the banal, the global and the everyday, and copresent them, making clear the connections and interdependencies of this new worldliness. The dream of the global conversation portrayed through the grainy live feeds of *Our World* is to be found here somewhere.

Journalists might then seek to embrace cosmopolitanism as a driving ethic, but this cannot happen without changes to editorial and commercial considerations which frame and dictate journalistic labour. Will the *New York Times* sack its foreign correspondents and replace them with local journalists who can give a more accurate picture of life beyond the US? Working for the *New York Times* is a prestigious gig which large numbers of elite educated Americans consider an acceptable and worthy career destination. As many foreign journalists are prone to do, they will regard themselves as archetypal cosmopolitans suited to the role and reject any suggestion that they do not understand conditions on the ground, while implicitly bringing their own bias to the table. Warnings from editors that foreign correspondents should avoid 'going native' expose the heart of this dissonance.

Anthropocene cosmopolitanism means recognising the Anthropocene cosmopolis. This is not a utopian place by any means; in the Anthropocene cosmopolis there are hierarchies of power and attempts to diminish the importance of what is going on elsewhere, but we can at least try to see it for what it is. As ever, this comes back to questions of work, of journalistic literacy and of editorial judgement. To write about the world at large, journalists must understand who they are and how they fit into it.

10

Global Journalists in Liquid Worlds

Questions of disintegration, fluidity, mobility and transience as the hallmarks of modernity have come to dominate how we see the future. Analysis and criticism of the post-1992 world have come to focus heavily on the interactions of media, individual choice, consumerism, financialisation, performance, self-identity and the new cosmologies of globalisation. While the future has always been a daunting place that promises and threatens at the same time, many of us find ourselves more preoccupied with holding our heads above water, or learning to live – in the words of Zygmunt Bauman (2000) – in liquid worlds.

Similar to Marx's ideas of constant disintegration and remaking, the idiom of liquidity has been the basis of critiques of the destabilisation of 'solid' modernity and its social costs. But it has also formed the foundation of managerialist and accelerationist approaches to dealing with social and economic shocks as mere collateral damage in the move from the present into the future. Indeed for those at the top, skin-deep readings of these new modernities have provided the cover for political and economic projects which worsen rather than alleviate the symptoms of an unstable present.

Bauman's central concept has also been applied to the present and future of journalism as a way of making sense of the inherent instability of both the sector and the world it operates in. This is a world where labour, economy, the environment and the self come together in quick and transient meetings. Change, flexibility and adaptation form the watchwords when both analysing liquid modernity and, in its most superficial forms, punting solutions to this transience and instability by getting rid of supposed blocks on liquid life in labour, finance and other norms to better help us navigate it. This facilitation by means of deregulation and cutting through the supposedly arbitrary restraints and inflexibilities of high modern social systems in terms of welfare, pensions and market caps has been clearly visible in the way journalism works too. Financial responsibility has been shifted from media organisations to their employees in the name of flexibility, and freelance labour has become a key component of journalistic work worldwide, even in formerly stable editorial roles.

When viewing the world with an Anthropocene gaze, however, liquidity is much more than a way of visualising economies and social life. The world is quite literally becoming more liquid as the global environmental system loses its stability and solidity, seemingly harder, faster and stronger every day. It is estimated that by 2100 sea levels will have risen between around half a metre and a metre in a best case scenario, while failure to reduce emissions could see them climb higher still (IPCC 2022). It is not just social systems but natural systems that are subject to quick and transient change. This means that liquidity is a question of the coproduction of climate, environment and social life, such are the implications of such huge climatic shifts for the orthodoxies of the contemporary world view.

The stories we tell about our own profession and the world at large are important in understanding, challenging and working with the reality of this liquidity. Ecomodernist journalism, for example – with a strongly deterministic bent and an enthusiasm for newness and changes as an end in itself – likes to present these changes as somehow historically necessary or inevitable (see, for example, Fahy and Nisbet 2017), integrating them within a broader narrative about modernity as a process of necessary evolution, disruption and adaptation. This is a story in which nuclear power, geoengineering and a turbocharged (and often nebulous) green innovation not only paper over the cracks of the current situation, but fit the whole temple of capitalism with some solar panels and a smart heating system. The present becomes nothing more than a test bed for human innovation that needs only to be unlocked and realised.

Critics of this approach can rightly accuse ecomodernist journalists of projecting the certainties of high 20th-century modernity and its steadfast belief in the future onto a situation which warrants confrontation with the very idea of progress. Indeed, since the early 1990s when the idea of a renewed project of modernity began to gain traction, with knock-on effects in the reformist left and globalisation movements, humanity is estimated to have emitted as much carbon as it had in the entirety of industrial history to that point, with levels roughly doubling between 1990 and the 2010s (Frumhoff, Heede and Oreskes 2015). These two parallel conceptions of journalism and social progress as a manifest journey conceive of the present as a difficult patch to be passed through, despite all evidence pointing to further instability and potential collapse along the line. We cannot, contrary to the most simplistic readings of the sociology of contemporary modernity, simply adapt by being footloose free thinkers and innovators. Literacy about these issues and the views embodied by journalists themselves is critical to cutting through the noise, but there is also a strong need to internalise this understanding as part of a reflexive self-awareness of the everyday conditions of being a journalist in the tides of a liquid world.

Thinking sociologically about liquid worlds

When we talk about 'liquid' journalism we are talking about journalism in a world 'where uncertainty, flux, change, conflict, and revolution are the permanent conditions of everyday life', as Mark Deuze (2008: 851) describes it. Both in freelance work and in larger organisations with salaried staff, we see a demand and a desire for multiskilled journalists who can cover multiple issues or multiple facets of an issue using a variety of platforms and technologies. An important component of this is the idea of the journalist not as technician or author, though these roles are still important, but also as curator in a world of infinite content, who must use their judgement to mediate the sea of information – what Deuze identifies as an interplay with the 'monitorial citizenship' of audiences who are not merely informed by reading 'the news' but who actively seek out ways to filter and keep a check on public life. In this regard liquid journalists are both monitorial citizens themselves and information points for monitorial behaviour on the part of their audiences.

Understanding liquid journalists is however much more than reimagining their role as technicians in a system of media labour. The information society as it now exists is much more banal than its early prophets imagined it to be, with its novelty morphing into resignation and desensitisation, but it is in the banalities of globalisation and social reorganisation that we can find answers about the journalists of the future.

Grasping the position of journalists in the global picture is a twofold undertaking. The first step is to look at the material conditions of their working life and the second is to consider their faculties as individuals somewhat divorced from the specifics of their employment, though these two things are at the same time closely related. In an interchangeable world of labour, journalism organisations are often less interested in who their employees actually are than what function they can fulfil, which establishes a reductive approach to the complexities of journalistic work and the institutional and shared memory needed.

To fully understand and convey climate change, a journalist might need to call on expertise and understanding not just of climate, but history, sociology, urban policy and data science. This need is at odds with the situation of many journalists today, forced to bear the costs of their own training but also rarely rewarded for bringing a more complex analysis of the situation to bear on their reporting. We cannot force all journalists to study for degrees in sustainability and the social sciences, but at a fundamental level there is a clear interest in making journalists ask – in Bauman's words – how their 'individual biographies intertwine with the history we share with other human beings' (Bauman and May 2019: 6).

Bauman claims that to think sociologically is to 'seek to make sense of the human condition via an analysis of the manifold webs of human interdependency – those toughest of realities to which we refer in order to explain our motives and the effects of their activation' (Bauman and May 2019: 6). This is similar to Michael Burawoy's (2005: 25) calls to view journalism as a form of public sociology, where social analysts need to cultivate a collaborative relationship between sociology and journalism to bring its gaze to bear on larger publics. The 'story' is not finite but ongoing and complex, and we can carry these broad frames of understanding with us wherever we go.

Among the repeated attempts at reform and declarations of journalistic self-image that vow to bring justice and social equity to the pages and screens of the public this might seem chimeric, yet journalism already engages in disciplining and conditioning through its economic and editorial structures. Training and practice is not an unmoving and pristine exercise, free from ideology and guided only by the hand of journalistic instinct and values and the pursuit of a story. Style guides and skills acquisition are not just a question of consistency of spelling and grammar, or of learning how to use equipment; they have deeper consequences for form and composition and the knowledge journalists create and propagate. Professional journalism programmes often make a point of their ability to educate journalists for the contemporary and future job market, but while some programmes place a heavy focus on understanding society and its entanglements as much as journalism itself, others merely see themselves as technical training for the next generation of news gatherers who can be slotted into jobs.

In Britain many elite journalists have no formal journalism education and begin working in the media straight out of university through well-established networks or traineeships. This can be contrasted to the situation in America where the grad school journalism programme has become a mainstay of media education, particularly those at elite universities such as Columbia which are financially prohibitive for many but claim to offer guarantees of a career in the top journalistic outlets of the English-speaking world.

Look to Sweden and the other Nordic countries and you will find a different culture again, with publicly funded journalism colleges that feed directly into relatively generously funded commercial and public media, or Russia and China where journalism education takes place in the context of limited public expression and with divergent traditions in terms of gathering and reporting news, including the understanding of what news is. At the heart of this diversity is the question of whether journalists are merely messengers and content producers or critical knowledge actors.

The conflict between journalism education as technical instruction and journalism education as a critical and interrogative practice is by no means new. In part due to the need to lend themselves academic legitimacy, journalism programmes have long sought to intellectualise the practice of journalism, with different degrees of success, either through invocation of its role as a public good and emphasising a particular ethic to journalism graduates, or through engaging in more critical reflection on the media more generally and hoping for the best once graduates are subject to the discursive and professional pressures of the industry.

Pushing such sociological lenses, and doing so reflexively so that journalists are self-aware about the frameworks of knowledge production they actively participate in, connects the banal everyday practices of news work with the scale and complexity of the world at large. Furthermore, how we interpret the entanglement of shared time and space and its relations is more complex even than looking at society alone. One of the significant bodies of knowledge to emerge from examination of the Anthropocene and the environmental crisis academically has been discussions of post-humanism and ecological realism, in which the relational human sociologies that usually preoccupy social scientists are extended to cover nonhuman participants. This is not necessarily out of a deep moral ethic as 'deep green' thinkers have suggested, but because of the recognition of the sheer importance of these agents to humanity, in the same way that tracing interhuman sociological relations seeks to understand how things *are* rather than how they should be. Looking back to Anna Tsing's expanded cosmopolitanism or Latour's actor–network theory, this means bringing the interdependent and coproductive systems that forge our realities into the domain of journalistic work.

Thinking sociologically in the Anthropocene has to look beyond narrow human relationships and – to borrow a journalistic turn of phrase – human interest stories. Instead we have to try and reconfigure our understanding of events and individual stories in terms of broader flows, fluctuations, integrations and disintegrations.

Seeing flow and disintegration

In a world that seems to be unravelling, melting slowly into a cycle devoid of clear markers and meanings, we should perhaps be less interested in specific things than in flows and processes, how they work and how we work within them. In the old systems of global media, foreign correspondents or subject-specific journalists were by definition tasked with providing the individual components of this bricolage, but were not in themselves responsible for the composite global readings of their work or for understanding the big ideas of historical unfolding in which we all live.

In liquid worlds this is more than just a straightforward question of representation. When journalism graduates and even more experienced reporters do encounter climate change and other Anthropocene shocks they may not know how to deal with it adequately, returning to the stock frames of politics, national borders and legitimate concerns. As Philo and Happer (2013) have shown in their applied fieldwork on future climate scenarios, how the press and by extension their audiences might react to future climate shocks and climate stories is not assured to be entirely constructive or positive. Using audience research and hypothetical news reports, one of their scenarios envisages 150,000 Bangladeshi refugees arriving on the south coast of England after catastrophic flooding, to be met by news crews, protestors and humanitarian support. If this were to come true and Europe were to be hit by waves of refugees that would make the migrant 'crisis' of the mid-2010s and 2020s pale into insignificance, who would be the journalists at different stages of the process contributing to the formation of this story? What would their education tell them to do, and how would they choose to contextualise the event? A journalist who has studied sociology or migration studies as part of their training might attempt to frame what was happening in a broader landscape, but what would the relationships between journalist, platform and audience be, and how would they be addressed? Furthermore, how does such a scenario play out in the context of balance when such events seem set to take place in a radicalised political climate?

In this story there are flows of people, water, money and political currents leading to a moment. Bangladesh, as East Pakistan and prior to that as part of India, was a British colony from which England and the rest of the 'core' countries of the British Empire benefited. Britain and its European neighbours were early industrialisers and the British Empire was key in the development of extractive industrial modernity around the world. Climate change has been caused by this process, so Britain owes a significant climate debt to emerging economies such as Bangladesh. This accounts for the moral explanation, but behind this there are also a series of more prosaic processes. Refugee conventions and welfare rights, housing pressures and migrant dispersion strategies that themselves have long trajectories are designed to regulate and facilitate the movement of people and capital. Understanding a situation like the one outlined by Philo and Happer requires an awareness of the inherent effects of these flows and their confluence.

The world in future will look very different, and what is unfolding in the media is merely the result of processes begun long ago, but the narrative thread in the previous paragraph is likely not one which most journalists would immediately spring to. The knee-jerk reaction of an editor might be that audiences would not understand it, or that it strayed dangerously close to being overly political in its hints at colonial restitution and 'rewriting' history as part of a larger volley in a culture war. The summary is, however, entirely based in historical and scientific fact. What we know about climate change invites us to look backwards and forwards at the same time. In liquid worlds, people are another flow alongside the broader unravelling of certainties and the often violent and overwhelming movements that follow. This logic is implementable in other contexts too – if journalism is to be both interrogative and generative it must operate from a knowledge base above and beyond that of media consumers, and it must find ways to implement that knowledge into the mainstream. As with the new cosmopolitanism, much of this potential – indeed the things that will make this possible – depends on the political economy of journalism and the media industries themselves.

The same flows and disintegrations that mark our understanding of the world at large are also evident in the fabric of media work and the creative economy. Similar to what David Hesmondhalgh and Sarah Baker (2011: 386) suggest, trying to accommodate understandings of creative and cultural labour in broad theories of modernity and global change threatens to be speculative and overgeneralised if done arbitrarily. At its most reductive this involves pointing to a universal object called 'capitalism' and assuming that all ills, big or small, stem from it,[1] yet making these links is necessary. To understand the stories we tell, we must link the mundane act of logging on for a shift or producing a commission to the macro-effects of modernity as part of a bricolage of events and processes. This also brings discussions of labour into broader conversations about the big questions of modernity and global change in journalism, and about how they move forward into the future together.

Journalism work as liquid life

Across the developed world the idea of the creative industries as an economic sector has emerged in tandem with the discussion of and enthusiasm for post-industrial knowledge economies, and it has expanded well beyond the arts as traditionally understood to include journalism and media work more broadly. One of the effects of this change is that journalism has moved from being institutionally anchored to more closely resembling other forms of creative work, not least in the idea of the journalist as a creative entrepreneur and as a personal brand in a marketplace of constantly competing news mouthpieces who must sell themselves and their products to publics. What labour academics have noted as the death of 'jobs' as an economic exchange between labour and employer, and their replacement by 'careers', is evident in modern media work. We constantly audition for new gigs, or even the same gig, rather than simply working in a more literal sense. This has become a moral project of continuous improvement which, in the words of the economic historian Alan McKinlay (2002), creates 'dead selves' held to account by the promise of future progression. The difference between classical accounts of careerism in solid bureaucracies and the world of liquid work is that the next promotion or professional advancement might come from a different set of managers and gatekeepers entirely, often in a shrinking market.

Such liquid practice has emerged partly as a result of technological changes specific to journalism and partly through the more general trend of neoliberal labour practices in which risks and responsibilities are transferred onto media workers. Liquid practice means media labour must constantly reapply for their own positions, where journalists can expect little loyalty from employers and are only as good as their next pitch or contract interview.

In this situation, finding positives to this 'brave new world of work' (Beck 2014) can be challenging, but in this destabilisation of creative labour we are also seeing the rise of decentralised networks of journalism and the emergence of new platforms for distributed media work in particular. The forced isolation of the COVID-19 pandemic in 2020 and 2021 appears to have exacerbated this, with newsrooms being compelled to work with staff who might as well be 2,000 km away as locked in the inner suburbs of South London or Abu Dhabi. Indeed parts of this book were written on a relatively remote peninsular on Scotland's west coast in the middle of a national lockdown using satellite broadband. Such digital resocialisation, in which workers are confronted with new workplaces and where journalists might spend more time with their subjects than with their colleagues, fundamentally changes the nature of journalistic networks, from a hub and spoke model to a looser constellation of correspondents and news gatherers with almost utopian potential.

This can all be seen in wider processes of disembedding in liquid modernity, in which we are removed from local contexts and interaction becomes distanced and more abstract. There has been significant research on how audiences become disembedded (see, for example, Widholm 2019) and the emergence of transnational audiences, but from the perspective of journalistic work the disembedding of journalists from strict physical spaces of work and their re-embedding in new contexts fundamentally changes conceptions of space and work for the journalists themselves.

A French journalist living in an apartment in Tokyo and able to work fluently in English, Japanese and French will touch any number of professional and personal identities and employers in the space of a few days. Like many international freelancers, their kit will consist of a laptop and probably a good-quality camera, as well as a podcasting microphone and a good enough broadband connection to do lives into news networks or prerecords at home. Apartments like this exist not just in Tokyo but in Edinburgh, Stockholm, Chicago, Mexico City, Buenos Aires, Vancouver, Shanghai, Harare – the list goes on. When discussing this disembedding it is easy to sound like an awestruck 1990s media studies major writing about the internet and its dreams of a decentralised utopia of users for the first time, but instead of living in cyberspace we live in physically disaggregated space.

Journalistic homes today and in the future can be compared (not entirely positively) to an artistic *atelier*, hybrid spaces of work and living, but also liminal spaces with one foot in their own street and another in a transnational network of creative practice. The hypothetical journalist in Tokyo might work for a journalistic startup based in Canada, or spend their morning putting together a short package for a broadcaster in Belgium and then do some lives into a studio at lunchtime for breakfast news elsewhere, then finish up by going to a local cafe to write a longer weekend feature (the public seem largely unaware that many of the 'correspondents' they see on global TV news are in fact freelancers wearing a network name badge). They might also go and pick up their kids or visit friends, take a walk in their local park and do some shopping, or sit down to work through their tax return. This typifies what Mark Deuze (2009: 471) describes as 'translocalism' in media production, but also the inherent instability of such practices, in which journalistic work is reduced to a series of finite interactions with different employers or audiences under a broader umbrella of media convergence. This media work takes place in the context of different types of media, for different markets and with different outcomes, from headline-grabbing audience retention to altruistic civic discussion.

Fundamentally, if we want to understand why journalists see and curate things in specific ways, we have to ask how the political economy of liquid media work determines how journalists experience the world. Media workers are not shielded from the realities of transnational markets or the inherent instability and precarity of the workforce more generally any more than their class allows. They have visa issues, struggle with childcare, become ill and have mortgages, and they live within modernity's structures just like the rest of us.

The ability to carry out such flexible work depends on having an economic baseline and diverse income streams, a situation which can lead to what the communications scholar Andrea Hunter (2016) has said is akin to having a second full-time job. Though it can be hard to find positives in this instability, the economic context also has the impact of potentially granting journalists significant agency through nontraditional sources of funding such as crowdfunding and sponsorship, engaging with different forms of audience and community. Economic security also comes through other forms of creative labour, including things such as translation, event moderation, copywriting and more generic content creation, which expand a journalistic skill set into less visible but potentially more financially lucrative areas. Liquid journalists do not just watch the world, but embody it as they seek to survive, for better or worse.

Living in liquid worlds

To be alive to the realities of liquid worlds, with their flows and disintegrations, their constant adjustments and realignments, we have to be open to and literate about everything that is going on around us and in our own lives. The problems that face liquid journalists are thus the problems that face the public at large under liquid modernity – banal and everyday confrontations with the labour market and the need to stay afloat, but also learning to live in a world that is already being transformed beyond recognition by the impacts of climate change, capital and technology.

Fundamentally, liquidity is a lens through which we can begin to grasp the undoing of the natural environment and of the certainties of solid modernity, its tropes, its expectations and its dreams. Drawing inspiration from the idea of journalism as modernity's textual system, we can also see the disintegration of these texts as part of the same broad flows, journalistic material and form being caught by the stream and remade. All that is solid is melting into air and water, and what that means for all of us, as both journalists and private citizens simply living today, can be terrifyingly uncertain.

For those who teach journalism or employ journalists there is also a reckoning with the future to be had that cannot simply be met with more innovation or footloose skills acquisition. The promise of linear careers and a life behind the headlines is now an empty one in many places, and so by encouraging journalists to be more reflexive about their own futures and their modes of practice we can encourage literacy about the world at large and the new and diverse forms journalism practice must necessarily take. Learning to live in this world, to understand why it is the way it is and how Anthropocene conditions change us all is crucial to finding a way out of it, on both the short timescales of stable careers and the long timescales of a stable climate.

Journalism still seems enraptured by the idea that it is somehow divorced from the politics and economics it reports on, but as the international freelancer with their curtains closed in a tropical heatwave will tell you, nothing could be further from the truth.

11

Journalism as Eschatology and the World to Come

> What is the meaning of these things unfolding before
> our eyes on the television?
>
> Ulrich Beck

Ultimately this is the problematic at the heart of journalism's encounter with the Anthropocene, not just on television but across multiple media. Ulrich Beck's exasperated question as he grappled with the overwhelming failure of the early 21st-century public to register what was in front of their eyes and the unravelling of his own understanding resonates through contemporary experience. We can all see these things unfolding, sometimes banal but often increasingly terrible, and yet making the connection to any coherent understanding of who we are and where we fit into these images remains chimeric.

This desire for meaning, or sense-making to use John Hartley's maxim, is what will define journalistic practice in the future and its usefulness as a distinct way of seeing and doing things in the broader sea of media, but it is also crucial to developing a shared structure of global understanding in which the basic narrative lines of the Anthropocene are established as foundational parts of being alive today.

Beck's words speak to a crisis of sociological insight as well as a crisis of media storytelling. As has been argued throughout this book, the risks and social bads which sociologists of modernity have identified as inherent to our understanding of the present have not been resolved, legislated or innovated away, but have instead become even larger existential threats which are now totalising in their extent. This has happened despite these threats being widely understood at a relatively early stage – we live in the shadow of an environmental wave in the 1990s and early 2000s that promised much but delivered little, and where the disconnect between consciousness and action is all too apparent. From Bill McKibben's backwoods epiphany to

Bruno Latour's declaration that we have never been modern, the holistic and interconnected turn of the 1990s and the rise of the Anthropocene in the following decade met the intellectual need to label the sensation of our environment transforming around us, yet we are still preoccupied with the future that should never have existed in the first place once the basics of climate science become commonly accepted knowledge.

It seems barely believable, but in the seven years since this book was first suggested as the basis for a research grant, climate change has begun to cut faster and deeper, and its immediate effects rather than its future risks have entered the public sphere in ways that were unthinkable only a decade ago. The 2023 IPCC report that set a 'code red' for humanity came off the back of forest fires and heatwaves around the northern hemisphere. The sight of the California hills ablaze and San Francisco – a city that has come to represent the hopes and nightmares of accelerationist tech capitalism – shrouded in impenetrable smoke, or the second lowest extent of Arctic sea ice on record, record bush fires in Australia, and new heat records across the world have all added to the immediacy of this encounter. In the middle stages of writing in 2022, London hit 40 degrees Celsius and homes burned as railway tracks buckled. A heat dome in North America led to the destruction of an entire town in British Columbia in a matter of hours, and a few weeks later flash flooding across western Europe killed hundreds and destroyed millions of euros worth of property. The Zackenberg Research Station in northern Greenland recorded its highest temperature to date of 24.3 degrees Celsius as global leaders dined in Glasgow to try and pledge to keep the average global temperature rise to 2 degrees, sidestepping the expert advice that the 1.5 degree aims of the 2015 Paris Agreement were not an immaterial political promise but a necessary ecological boundary.

Beyond climate change, the markers of the Anthropocene continue to stack up in less tangible ways. At the earth's deepest point, the Mariana Trench, microplastics swirl in the currents of the ocean floor; they are frozen in the Antarctic ice and even found in human placentas and in clouds. We don't just live within modernity; modernity's material traces now live within us. We have less become one with nature than with the toxic legacy of the world system we have created.

No doubt further heat records will be broken and new thresholds passed in the months and years immediately after these words are put out into the world. They will date and need updating; when the solutions to the Anthropocene have to be implemented within the next decade or so to avoid further catastrophic damage, the time frames of book publishing will now overlap with the time frames of mass system change. This is earth history in real time but also in media time.

Climate change is the thing that threatens us most acutely and occupies the headlines, but the phosphate build-ups, microplastics, soil degradation and biodiversity loss are all symptoms of a present that will be with us for some time. As capital accumulates, the inequality gap continues to grow. This is not about management of a single issue or awareness raising as part of a risk-based solutionism, but about modernity and capitalism as media objects, and of how we understand their effects and our own place in them through journalism.

This does not mean playing into the hands of those pushing business as usual in journalism or politics. The paternalism of global management that claims to recognise and resolve problems in sensible ways is demonstrably not working. Being aware of capitalism, modernity and their entanglements is integral to understanding the possibilities of more productive futures and the processes of the past that have formed the present, yet the misrepresentation of these critiques also risks becoming fuel for the culture wars that trade on lazy stereotypes of alternative modes of organisation. In much the same way that the political system at large struggles to articulate alternatives to the status quo despite the obvious need, journalism is still struggling to make sense of itself, and continues to suffer as a result. We are very aware of the problems in the system but remain reluctant to remove the toxic elements that study after study shows to be there.

The question that cuts through the centre of this is the visibility of these systems and the realities that underpin them to the audiences and publics who experience their effects. This bind has been described by the Canadian journalism scholars Candis Callison and Mary Lynn Young (2019) as a need to locate journalism not just in terms of its own institutional crisis, but in terms of its potential as a form of discursive repair for the wider world. Beyond the communities of academic practice and environmental science, the existential implications of Anthropocene thinking are hard to discern. The Anthropocene is not a widespread or visible concept and it will likely never be so. Journalism, meanwhile, in its normativity labours on, trying to recreate a sense of what it perceives to be normal. This is not just a matter of seeing journalism as a 'zombie institution' as many have suggested, but of the world it reports as a form of 'zombie modernity' which shuffles along as a pallid reflection of something vibrant and living but which is in fact half dead. This gap between perception and reality pervades the future of journalism as it grapples with the intrusion of Anthropocene effects and economic realities into its everyday, and ultimately asks us to go about locating some kind of meaningful reality beyond the purely representational – to get real about the processes behind everyday life.

Getting real about the Anthropocene

This dissonance calls into question journalism's ability to function as a representative practice and its core characterisation as a sense-making function in modernity. Journalists and audiences alike need to look beyond 'the story' and relay what is actually going on, however much that might jar with established genres and conventions. Borrowing from the extensive work done on nonrepresentational theory carried out in cultural geography, it becomes clear how reporting the Anthropocene and placing it within journalism is, in the words of the geographer Hayden Lorimer (2005), a 'more than representational' undertaking. The filtration, and the reductive streamlining of narratives which ignores both contemporary experience and the complexity of the systems in which these experiences happen, mean that journalism's hunt for a good story often results in a fundamentally misleading presentation of events and connections.

Following the line of nonrepresentational thinking, the American journalism scholar Perry Parks describes news happenings in terms of what the historian Daniel Boorstin termed 'pseudo events' (Parks 2017b), which fail to take into account the multitude of other things happening which are of relevance and importance to the public. These happenings – press conferences, unveilings, the issuing of reports or speeches by politicians and stakeholders – are all attempts at representing *something*, but these events are not always linked to any cogent explanation of why these things are happening. This critique was developed in the cut and thrust of American media life and its thirst for action, an internalised compulsion for staged realities and projections, but it applies more broadly to journalism's world view and its representational and editorial instincts.

A 'more than representational' mindset, according to Parks (2017b: 126), would 'pay close attention to the pantheon of influences occurring right now that affect and make possible what will happen next'. Journalism about the future, about the Anthropocene and its trajectory, has to deal with this material basis to all it seeks to report or imagine, and pay attention to the things happening right now which either create or deny possibilities for events to happen in the future. Realism, however, is often taken to mean being realistic about what is politically feasible or what can conform to existing genres of reporting and political action. Reporters at climate summits veer towards adopting a form of political realism in which national actors are more important than transnational processes (Kunelius and Eide 2012: 277), reproducing established norms of political journalism at a moment of transnational opportunity by taking an interest less in the overall fate of the planet than in the political horse-trading of individual delegations.

At the Glasgow climate summit, the biggest media scrum was for a clear shot of Barack Obama walking down some stairs. This fixation on the political took place in tandem with an approach to targets and reforms which were considered to be realistic and reasonable, within the envelope of what diplomacy and political imagination could envisage. Climate and Anthropocene journalism is thus pulled in different directions as it seeks to balance the obvious need for transformative social change with the realism of the international and national political arena and its power brokers, ideological limits and modes of communication.

At the heart of this compulsion to be realistic about climate change and other Anthropocene problems – and the rejection of utopian solutions in the face of the goings on of global climate politics – is a false pragmatism. Indeed the ethic of realism which dictates government policy on climate change, but also on other things critical to our wellbeing such as protections for biodiversity and classical concerns such as the use of phosphates and pesticides, is easy to find. Neither domestic nor international politics is prepared to think the unthinkable in terms of solutions, though it is prepared to let the unthinkable happen in terms of climate damage. This is what Joel Wainright and Geoff Mann (2013) refer to as a 'climate leviathan' in which global governance adapts to climate change without fully recognising the implications of what has happened. The early 2020s heatwaves in Europe, for example, saw concern about record heat and climate quickly flip into a discussion of whether it was realistic to channel public money to energy demand reduction, and whether voters would 'buy' it at a time of skyrocketing energy bills.

As Wainwright and Mann summarise it, what seems likely is that a commitment to remaining realistic and coming up with 'workable' and acceptable solutions merely creates a model of governance which maintains ultimately detrimental broad structures of government and social organisation. Exasperated journalists might well notice the similarities to the innovation over revolution mantra of media companies, and the 'realism' of the leviathan which normalises 'Trading in carbon-emissions permits ("cap-and-trade"), judicious market assessments of "tradeoffs", nuclear power, corporate leadership, carbon capture and storage, green finance, and ultimately, geoengineering' (Wainwright and Mann 2013: 6) The arguments will be familiar to anyone who has been assured that their employer is looking at ways it can support quality journalism and grow audiences, including equally judicious assertions of their organisation's role for the public good – the leviathan is the body politic of political economy that refuses to accept its own ultimately problematic nature.

In this scenario, realism means doing everything apart from what is materially necessary because that would be deemed unrealistic. In this world of anything-buts we can entertain any number of what-ifs, trading on intention as much as action and the idea that the future is a malleable concept that can be rendered concrete based on whatever discourse enters the news media. Headlines about machines that pull carbon from the air, adding ethanol to petroleum to make cars greener, and lifestyle articles about sustainable ecobreaks create a future which is reassuring in its belief that modernity merely requires some technological tweaking around the edges and consumer behaviour changes to get back on track, even when all of the science points to a very different reality with severe implications for the practice of everyday life.

Breaking out of representation is not easily done – editors and reporters are as exposed to the ups and downs of media discourse as the public, even when they may themselves see their work as an exercise in reverent objectivity. Media gatekeeping means that the complex interactions of audiences, reporters, commercial concerns and editorial lines can lead to the triumph of pseudo-events over engagement with what is actually happening, and what will happen. This also challenges preconceived notions about resolution, making it hard to offer neat packages of the future when it is so open ended and potentially catastrophic, rather than conclusive and well behaved.

Journalism is not a pristine practice that exists in isolation – it needs audiences and it needs a basis on which to operate. Calls for journalism to 'do better' or to paternalistically feed its publics the correct information about what is to be done are ultimately deeply naive, and realism for journalism means realism about what it can and cannot do as well as the basic truth value statements of its coverage. A long read on climate change and modernity in the *New York Review of Books* might command critical acclaim and approval from its readership but is unlikely to set the world on fire, even when large parts of the world are in fact on fire. Getting real about Anthropocene media also means tapping into wider social currents and mass media, and thinking critically about the political economy of media as much as its content and journalistic practice.

Journalism, climate and the sociotechnical imaginary

The future, then, is uncertain, if not entirely unknowable. The solutions offered for both journalism's malaise and the future of the planet are a combination of social change and technological adaptation and restraint, and there is no doubt that both will be needed. Trying to find some kind of stability for the environment and our ways of mediating in a liquid world ultimately means making peace with modernity's impact on both strands of the problem.

This can be framed using the Indian science and technology theorist Sheila Jasanoff's (2015: 24) concept of the sociotechnical imaginary, or 'shared understandings of forms of social life and social order attainable through, and supportive of, advances in science and technology'. These shared understandings of the coproduction of social life and technology are critical to finding ways through the future and of critically understanding what we are going through. Just as the wilderness fantasies of deep ecology hark back to a time of pristine environment, the news industry's desire to recreate an idealised past version of itself and its values stands in sharp denial of the sheer inevitability of change. As Jasanoff points out, at times the 'juggernaut of global capital, driven by the furious whip of technological innovation and tuned to a univocal discourse of progress, seems unstoppable' (2015: 339). This inevitability, however, is part of the imaginary too. These dreamscapes (and nightmares) of modernity appeal to a particular narrative predilection for grand visions, albeit socially shared and embedded. This is where the idea of journalism as modernity's primary sense-making practice gains new currency. Journalism has a role to play in constituting and guiding future imaginaries, but it is also subject to the same processes of future building through its own need for fruitful imaginaries to regain internal stability and its own nominal place as the foremost sense-making practice.

Such imaginaries are useful because of how they inform decisions made in the present, irrespective of how real the worlds to come actually are. These future imaginaries help to formulate and reformulate the future as it is anticipated. Such imaginaries are value-laden, from the bright clean cities of urban development's CGI renderings to the overgrown ruins of radical environmentalism's Ruritanian post-capitalist utopias. Each invites us to briefly live in the future and then return to the present to set about making it happen, and to establish a consensus about what is possible and desirable.

Ultimately, these imaginaries help to constitute the delicate agency of the present as much as scientific discourse does, especially in the popular media space. This agency is delicate because it is neither fully formed nor sustainable, and as each day passes the ability to do some things diminishes and the imperative to do others increases. We know that climate change is already well underway and that even if its causes were frozen tomorrow it could continue to have impacts well into this century. We are also now living with the realities of climate disruption on an annual basis and increasingly well-modelled scenarios of what this means going forward, and we know that the prevalence of Anthropocene processes across the globe will continue to have impacts over multiple lifetimes. Making sense of the present is a question of understanding this reflexivity and the quick and slow modes of time that overlap and intermingle in its wake.

The delicate agency of journalism, too, and the concept of journalists and journalistic institutions as agents of change depend heavily on this realisation. As journalists grapple with post-institutional work, the old questions of power and agency are rewritten. Instead journalism increasingly comes to manifest as a series of micro-actions and participatory interactions. Journalism itself cannot change the world, but Anthropocene journalism might at least begin to offer a version of reality anchored not only in science, but in collective democratic experience and material political fact.

It is also here that we can see the dissonance inherent to media in mapping the future. If, as Mike Gasher (2015) suggests, we can indeed see journalism as a process of mapmaking, this creates expectations that something is just around the corner, only to find that thing in a very different state to what we had been led to believe. Right now it is critical that we can see exactly where we are and which routes are available to us – in climate and environmental science, this is the collective burden of the IPCC and its attempts to map the future and identify pathways for relative prosperity in the circumstances. The journalistic geographies alongside this are tasked with their own kind of mapmaking, of representing what this future will be like in order to find agency in the present.

This delicate agency is mirrored in Barbara Adam's (2018: 388) maxim, 'What can the future do for us? What are we doing to the future?' These two questions operate in tandem with one another, as the answer to the first is dependent on the circumstantial factors arising from the second. Quite simply, by acting in certain ways in the present we either facilitate or circumscribe the possibilities of the future. The present moment and the present moments that will follow it also embody a potential not to reset and regain what is being lost each minute and second, but to move through the Anthropocene and to keep going. This can be combined with a realism that does not underestimate the task of what is ahead. The challenge is to be hopeful but not naive, or to embrace realism without despair. While a new electric car scheme or children planting trees at a community nature reserve are objectively good news, they are not themselves solutions to the problems at hand.

This is evident in journalism's own futurological crisis too, where a successful startup running for one year or hiring a single extra employee is framed as a success, even when experience counts against the idea of ventures and novel technologies succeeding. The trend in journalism is decidedly towards a further hollowing out of its institutions and analytical capacity. At the same time the idea of total failure of the journalistic mode is unthinkable and still unlikely. In a world that wants magic bullets and disruption, game changers in technology and society, the assertion of our own limitations is a sobering act. This limit to agency, indeed the recognition of how compromised we are and how limited the idea of a quick solution to the complex challenges of modernity and its impacts is, promotes a more constructive appraisal than might otherwise be possible.

In an Anthropocene context this fragile situation operates in a fine balance between the past and many possible futures. As journalists try to find ways of conveying the scale of what is happening and will continue to happen, it might appear that the journalist as the guiding hand of sense-making should hand in their laptop and hand over the public space to the social media output of campaign groups, lobbyists and the general public. In this moment, though, journalists and journalism are faced with a choice – organise to resist the effects of liquid media capitalism or try to internalise its values and adapt to its fluid rulebook in the name of innovation and good business.

This dilemma is the reason the political economy of journalism is central to its imaginative and generative capacity; in a global situation where the needs of all of us are often dramatically underserved by the existing infrastructures and economies of media, questions of media 'resilience' or innovation are often ways of failing to discuss how and why journalism has reached its present impasse. It does not exist in an external sphere untrammelled by the movement of the world at large, but is constitutive of and is constituted by this environment. Ultimately it comes down to an acceptance of the fact that existing economies – in a manifestation of Lauren Berlant's *cruel optimism* – undermine their own promise. Things which are set up as linear tracks for future progress return in unwanted forms and what seem like solutions return as obstructions. Futures which can be more than speculative must also be realist.

Eschatology and the end of now

In this delicate present we can begin to characterise journalism – if it takes its role of documenting the present and forming the future seriously – as an eschatological process (see Hinde 2021). Eschatologies are not merely the stories of endings, but stories of transitions to dramatically new circumstances. They are also tales of deep irony, what the novelist Amitav Ghosh (2018: 32) describes as 'the mysterious work of our own hands returning to haunt us in unthinkable shapes and forms'. The end times of journalism may have become a trope in the contemporary discussions of journalistic futures, but we can also characterise all journalism as a collective genre of eschatology as it tries to cope with the unravelling of the neat futures of modernity.

This might sound a touch melodramatic, but what we know for certain is that the future is going to be extremely different, and that this will have a severe impact on the things journalism seeks to cover, and by extension on journalism itself. To understand this, seeing the climate crisis and the crisis of journalism as facets of the same processes is critical, not least in how they both relate to and represent and constitute effects of modernity, and consequently how reimagining modes of organisation is critical to positive outcomes for both.

While it is dangerous to forecast the finality and close of anything – we need only look at the hubris of the 1990s and its liberal utopias – the sheer force of climate change in particular means that we can be certain that both our immediate and medium futures are going to be substantially altered. This will likely not merely be a disruption but a total reconfiguration of modernity as we know and experience it, both in the consumption and production of media and in the events of our own daily lives.

This acceptance of the inevitability of ending – what Roy Scranton (2015) calls 'learning to die' in the Anthropocene – is crucial to understanding what comes after. The question we face is whether we can assemble institutional and democratic support for relatively positive outcomes in the face of overwhelming evidence that our present ones are not fit for purpose. As this applies to journalism, this means seeing the present media crisis not as a disruption but as a remaking. In these times, journalism is also a form of public sociology, a 'searching for order in the broken fragments of modernity' (Burawoy 2005: 5) which can use 'extensive research, elaborate theories, practical interventions, and critical thinking, reaching common understandings across multiple boundaries' (Burawoy 2005: 25). Journalism can reach through the social world, and fundamentally make sense of it from its fragmentary dissonance. If journalism is to maintain its position as the sense-making practice of modernity, it must then also function as a hand-holding practice for the end of one epoch and the beginning of another. The material evidence for global change is so overwhelming that journalism can no longer see itself as an early warning system – journalism in the immediate and medium future, even under positive outcomes, will be a practice of hand-holding and sense-making for a world which may well be scary and unfamiliar, a world in which democratic and civic institutions are not guaranteed to hold up, and in which new forms of media practice and ways of seeing will come to the fore.

The Anthropocene idea is characterised by its discontinuity. This is not a staging post in the onward march of human history but a chastening. In the procedural world of news media and its shifting time horizons, eschatology asks us to think differently about how we portray the present and its narrative betterment. Journalism is an ecosystem service, a public good and an object of utopian utility. Finding solidity in instability, or forming structure from collapse, is the ultimate challenge presented by the oncoming future. As we wade into the world to come, we must confront media as part of the modernity we have created and modernity as propelled and fed by media, finding strength in our ability to see more generative futures than those currently unfolding on the screens in front of us.

Notes

Chapter 2
[1] In 2020 thousands of songbirds were documented falling from the sky in the southern United States, with studies showing they were emaciated and impacted by climate change-induced cold temperatures (Weston 2020).

Chapter 3
[1] In the early years of fossil transport, rail accidents were used to evoke the violence of modernity as previously unthinkable speeds and collisions became commonplace. Writers Charles Dickens and Gerhart Hauptmann both penned short stories on the potential horror of the railways, and Dickens was involved in an early rail accident that killed several people.

Chapter 5
[1] As the French historian of science Alain Desrosières (1998) has argued, large-scale data use has been integral to modern and premodern states since well before the digital revolution.
[2] Anti-rationalist and select works of postmodern philosophy have gained popularity among conservatives seeking to challenge scientific knowledge and consensus by casting themselves as critical free thinkers, not prisoner to groupthink and political bias.
[3] Exxon, for example, are widely believed to have held relatively detailed internal data on climate change in the late 1980s for the purposes of market planning, long before it was made available to international policy makers or confirmed through the work of the IPCC.

Chapter 6
[1] Glasgow takes on exactly this role in Kim Stanley Robinson's near future realist sci-fi novel *The Ministry for the Future*.
[2] This is based on calculations factoring the total production in MW per year relative to annual consumption in MW.

Chapter 9
[1] In many cases the 'chance' to appear on broadcast networks is presented as payment in itself for the journalistic knowledge presented.

Chapter 10
[1] In this regard, using modernity as a macroframe is useful, because it contains capitalism as a major constituent part but is open to pluralistic approaches that look at effects beyond simple readings of capitalism as a uniform object over time and space.

References

Abbinnett, Ross. 2019. The Anthropocene as a figure of neoliberal hegemony, *Social Epistemology* 33 (4): 367379.
Adam, Barbara. 1998. *Timescapes of Modernity: The Environment and Invisible Hazards*. Routledge.
Adam, Barbara. 2018. Four meditations on time and future relations, *Time & Society* 27 (3): 384–8.
Aiello, G., Tarantino, Matteo and Oakley, Kate (eds). 2017. *Communicating the City: Meanings, Practices, Interactions*. Peter Lang.
Alexievich, Svetlana. 2016. *Chernobyl Prayer: A Chronicle of the Future*, trans. Anna Guinin and A.L. Tait. Penguin Press. Originally published in Russian in 1997.
Amin, A. and N. Thrift. 2016. *Seeing Like a City*. Polity Press.
Anderson, Ben. 2015. What kind of thing is resilience?, *Politics* 35 (1): 60–6.
Anderson, Benedict R.O.G. 1983. *Imagined Communities: Reflections on the Origin and Spread of Nationalism*. Verso Books.
Anderson, C.W. 2018. *Apostles of Certainty: Data Journalism and the Politics of Doubt*. Oxford University Press.
Baldwin, Andrew, Christiane Fröhlich and Delf Rothe. 2019. From climate migration to Anthropocene mobilities: Shifting the debate, *Mobilities*, 14 (3): 289–97.
Barney, Darin, Gabriella Coleman, Christine Ross, Jonathan Sterne and Tamar Tembeck (eds). 2016. *The Participatory Condition in the Digital Age*. University of Minnesota Press.
Bauman, Zygmunt. 2000. *Liquid Modernity*. Polity Press.
Bauman, Zygmunt and Tim May. 2019. *Thinking Sociologically*. John Wiley & Sons.
Beck, Ulrich. 1992. From industrial society to the risk society: Questions of survival, social structure and ecological enlightenment, *Theory, Culture & Society* 9 (1): 97–123.
Beck, Ulrich. 2002. The cosmopolitan society and its enemies: Special issue on cosmopolis, *Theory, Culture & Society* 19 (1–2): 17–44.
Beck, Ulrich. 2006. *The Cosmopolitan Vision*. Polity Press.
Beck, Ulrich. 2014. *The Brave New World of Work*. John Wiley & Sons.

Bell, Karen. 2020. *Working-Class Environmentalism: An Agenda for a Just and Fair Transition to Sustainability*. Springer Nature.

Berglez, Peter. 2008. What is global journalism? Theoretical and empirical conceptualisations, *Journalism Studies* 9 (6): 845–58.

Berlant, Lauren Gail. 2011. *Cruel Optimism*. Duke University Press.

Berman, Marshall. 1983. *All that is Solid Melts into Air: The Experience of Modernity*. Verso Books.

Berry, Mike. 2019. *The Media, the Public and the Great Financial Crisis*. Springer International Publishing.

Bhambra, Gurminder K. 2016. Whither Europe? Postcolonial versus neocolonial cosmopolitanism, *Interventions* 18 (2): 187–202.

Bogaerts, Jo and Nico Carpentier. 2013. The postmodern challenge to journalism: Strategies for constructing a trustworthy identity, in Chris Peters and Marcel Jeroen Broersma (eds), *Rethinking Journalism: Trust and Participation in a Transformed News Landscape*. Routledge, pp 60–71.

Boyer, Dominic. 2013. *The Life Informatic: Newsmaking in the Digital Era*. Cornell University Press.

Bregman, Rutger. 2018. *Utopia for Realists*. Bloomsbury Publishing.

Brown, Kate. 2020. *Manual for Survival: A Chernobyl Guide to the Future*. W. W. Norton & Company.

Bulletin of the Atomic Scientists. 2020. Press release – It is now 100 seconds to midnight. Bulletin of the Atomic Scientists, 23 January. Available at: https://thebulletin.org/2020/01/press-release-it-is-now-100-seconds-to-midnight/ [Accessed 20 May 2024].

Burawoy, Michael. 2005. For public sociology, *American Sociological Review* 70 (1): 4–28.

Buzzanell, Patrice M. and J. Brian Houston. 2018. Communication and resilience: Multilevel applications and insights – a *Journal of Applied Communication Research* forum, *Journal of Applied Communication Research* 46 (1): 1–4.

Calhoun, Craig. 2008. Cosmopolitanism in the modern social imaginary, *Daedalus* 137 (3): 105114.

Callison, Candis and Mary Lynn Young. 2019. *Reckoning: Journalism's Limits and Possibilities*. Oxford University Press.

Carson, Rachel. 1962. *Silent Spring*. Penguin Books.

Caswell, David. 2016. Reporting into structure: How journalists, crowds and robots can work together. Donald W. Reynolds Journalism Institute, 26 January. Available at: https://rjionline.org/reporting/reporting-into-structure-how-journalists-crowds-and-robots-can-work-together/ [Accessed 22 September 2025].

Christophers, Brett. 2023. *Our Lives in Their Portfolios: Why Asset Managers Own the World*. Verso Books.

Coddington, Mark. 2015. Clarifying journalism's quantitative turn: A typology for evaluating data journalism, computational journalism, and computer-assisted reporting, *Digital Journalism* 3 (3): 331–48.

Cottle, Simon. 2011. Taking global crises in the news seriously: notes from the dark side of globalization, *Global Media and Communication* 7 (2): 77–95.

Cronon, William (ed). 1996. *Uncommon Ground: Rethinking the Human Place in Nature*. W.W. Norton & Company.

Crutzen, Paul and Eugene Stoermer. 2000. The 'Anthropocene', *IGBP Newsletter* 41: 17–18.

Daniels, Jessie, Karen Gregory and Tressie McMillan Cottom (eds). 2016. *Digital Sociologies*. Policy Press.

Delanty, Gerard. 2009. *The Cosmopolitan Imagination: The Renewal of Critical Social Theory*. Cambridge University Press.

Demos, T.J. 2017. *Against the Anthropocene:* Visual Culture and Environment Today. Sternberg Press.

Desrosières, Alain. 1998. *The Politics of Large Numbers: A History of Statistical Reasoning*. Harvard University Press.

Deuze, Mark. 2006. Participation, remediation, bricolage: Considering principal components of a digital culture, *Information Society* 22 (2): 63–75.

Deuze, Mark. 2008. The changing context of news work: Liquid journalism for a monitorial citizenry, *International Journal of Communication* 2 (5): 848–65.

Deuze, Mark. 2009. Media industries, work and life, *European Journal of Communication* 24 (4): 467–80.

Deuze, Mark. 2012. *Media Life*. Polity Press.

Deuze, Mark. 2019. What journalism is (not), *Social Media + Society* 5 (3): 205630511985720.

Donovan, Kevin P. 2020. Colonizing the future, *Boston Review*, 28 September. Available at: www.bostonreview.net/articles/colonizing-the-future/ [Accessed 22 September 2025].

Dryzek, John S. 2016. Institutions for the Anthropocene: Governance in a changing earth system, *British Journal of Political Science* 46 (4): 937–56.

Duarte, Carlos M., Lucille Chapuis, Shaun P. Collin, Daniel P. Costa, Reny P. Devassy, Victor M. Eguiluz, Christine Erbe, et al. 2021. The soundscape of the Anthropocene ocean, *Science (American Association for the Advancement of Science)* 371 (6529): eaba4658.

Edwards, Paul N. 2017. Knowledge infrastructures for the Anthropocene, *Anthropocene Review* 4 (1): 34–43.

Elhacham, Emily, Liad Ben-Uri, Jonathan Grozovksu, Yinon M. Bar-On and Ron Milo. 2020. Global human-made mass exceeds all living biomass, *Nature* 588 (7838): 442–4.

Fahy, Declan and Matthew C. Nisbet 2017. The ecomodernists: journalists reimagining a sustainable future, in Peter Berglez, Ulrika Olausson and Mats Ots (eds), *What is Sustainable Journalism?: Integrating the Environmental, Social, and Economic Challenges of Journalism*. Peter Lang, pp 39–57.

Farrier, David. 2020. *Footprints: In Search of Future Fossils*. 4th Estate.

Fast, Karin, Henrik Örnebring and Michael Karlsson. 2016. Metaphors of free labor: A typology of unpaid work in the media sector, *Media, Culture & Society* 38 (7): 963978.

Frumhoff, Peter C., Richard Heede and Naomi Oreskes. 2015. The climate responsibilities of industrial carbon producers, *Climatic Change* 132 (2): 157–171.

Fukuyama, Francis. 1989. The end of history?, *National Interest* 16: 3–18.

Garnett, Philip. 2018. Total systemic failure?, *Science of the Total Environment*, 626: 684–8.

Gasher, Mike. 2015. Geographies of the news, in Susan P. Mains, Julie Cupples and Chris Lukinbeal (eds), *Mediated Geographies and Geographies of Media*. Springer, pp 127–40.

Houston, J. Brian, Megan K. Schraedley, Mary E. Worley, Katherine Reed and Janet Saidi. 2019. Disaster journalism: fostering citizen and community disaster mitigation, preparedness, response, recovery, and resilience across the disaster cycle. *Disasters*, 43 (3): 591–611.

Gabrys, Jennifer. 2022. *Citizens of Worlds: Open-air Toolkits for Environmental Struggle*. University of Minnesota Press.

Gabrys, Jennifer, Michelle Westerlaken, Danilo Urzedo, Max Ritts and Trishant Simlai. 2022. Reworking the political in digital forests: the cosmopolitics of socio-technical worlds, *Progress in Environmental Geography* 1 (1–4): 58–83.

Ghosh, Amitav. 2016. *The Great Derangement: Climate Change and the Unthinkable*. University of Chicago Press.

Gibbs, Ewan. 2021. Scotland's faltering green industrial revolution, *Political Quarterly* 92 (1): 57–65.

Gibson, Timothy A., Richard T. Craig, Allison C. Harper and Jordan M. Alpert. 2016. Covering global warming in dubious times: Environmental reporters in the new media ecosystem, *Journalism* 17 (4): 417–34.

Gössling, Stefan and Andreas Humpe. 2020. The global scale, distribution and growth of aviation: Implications for climate change, *Global Environmental Change* 65: 102194.

Hamilton, Clive. 2016. The theodicy of the 'good Anthropocene', *Environmental Humanities* 7 (1): 233–8.

Hannerz, Ulf. 2004. *Foreign News: Exploring the World of Foreign Correspondents*. University of Chicago Press.

Happer, Catherine. 2017. Financialisaton, media and social change, *New Political Economy* 22 (4): 437–49.

Happer, Catherine. 2024. *The Construction of Public Opinion in a Digital Age*. Manchester University Press.

Haraway, Donna. 2015. Anthropocene, capitalocene, plantationocene, chthulucene: Making kin, *Environmental Humanities* 6 (1): 159–65.

Harcup, Tony and Deirdre O'Neill. 2017. What is news? News values revisited (again), *Journalism Studies* 18 (12): 1470–88.

Hartley, John. 1996. *Popular Reality: Journalism, Modernity, Popular Culture*. Arnold.

Hartley, John. 2008. Journalism as a human right: The cultural approach to journalism, in Martin Löffelholz and David Weaver (eds), *Global Journalism Research: Theories, Methods, Findings, Future*. Peter Lang, pp 39–51.

Hesmondhalgh, David. 2021. Is music streaming bad for musicians? Problems of evidence and argument, *New Media & Society* 23 (12): 3593–615.

Hesmondhalgh, David and Sarah Baker. 2011. *Creative Labour: Media Work in Three Cultural Industries*. Routledge.

Hinde, Dominic. 2021. Travelling in hope, *Sunday Post*, 12 November.

Horne, Mark. 2020. Climate change warning after Scottish rail landslide, *Times Scotland*, 31 August.

Hunter, Andrea. 2016. 'It's like having a second full-time job': Crowdfunding, journalism and labour, *Journalism Practice* 10 (2): 217–32.

IFRC. 2021. *Displacement in a Changing Climate: Localised Humanitarian Action at the Forefront of the Climate Crisis*. IFRC.

Ingold, Tim. 2010. Footprints through the weather-world: Walking, breathing, knowing, *Journal of the Royal Anthropological Institute* 16 (s1): S121–39.

IPCC. 2022. *Climate Change 2022: Impacts, Adaptation, and Vulnerability: Contribution of Working Group II to the Sixth Assessment Report of the Intergovernmental Panel on Climate Change* [H.-O. Pörtner, D.C. Roberts, M. Tignor, E.S. Poloczanska, K. Mintenbeck, A. Alegría, M. Craig, S. Langsdorf, S. Löschke, V. Möller, A. Okem and B. Rama (eds)]. Cambridge University Press.

IPCC 2023. *Climate Change 2023: Synthesis Report. Contribution of Working Groups I, II and III to the Sixth Assessment Report of the Intergovernmental Panel on Climate Change*. IPCC.

Jacobs, J. 2007. *Dark Age Ahead*. Vintage.

Jaeger, Carlo C. and Julia Jaeger. 2011. Three views of two degrees, *Regional Environmental Change* 11: 15–26.

Jamie, Kathleen. 2008. A lone enraptured male, *London Review of Books*, 30 (6).

Jasanoff, Sheila. 2015. Imagined and invented worlds, in Sheila Jasanoff and Sang-Hyun Kim (eds), *Dreamscapes of Modernity: Sociotechnical Imaginaries and the Fabrication of Power*. University of Chicago Press, pp 321–41.

Kanngieser, Anja. 2015. Geopolitics and the Anthropocene: Five propositions for sound, *GeoHumanities* 1 (1): 80–5.

Keys, Patrick, Victor Galaz, Michelle Dyer, Nathanial Matthews, Carl Golke, Magnus Nyström and Sarah E. Cornell. 2019. Anthropocene risk, *Nature Sustainability* 2 (8): 667–73.

Kitch, Carolyn L. 2006. 'Useful Memory' in Time Inc. Magazine: Summary journalism and the popular construction of history, *Journalism Studies* 7 (1): 94–110.

Koch, Alexander, Chris Brierley, Mark M. Maslin and Simon L. Lewis. 2019. Earth system impacts of the European arrival and Great Dying in the Americas after 1492, *Quaternary Science Reviews* 207: 13–36.

Kolbert, Elizabeth and David Roberts. 2017. Covering catastrophe. Climate One. Available at: www.climateone.org/audio/elizabeth-kolbert-and-david-roberts-covering-catastrophe.

Kunelius, Risto and Elisabeth Eide. 2012. Moment of hope, mode of realism: On the dynamics of a transnational journalistic field during UN climate change summits, *International Journal of Communication* 6 (1): 266 –85.

Langston, N. 2010. *Toxic Bodies: Hormone Disruptors and the Legacy of DES.* Yale University Press.

Latour, Bruno. 2012. *We Have Never Been Modern.* Harvard University Press.

Latour, Bruno. 2016. Fifty shades of green, *Environmental Humanities* 7 (1): 219–25.

Lazarus, Richard J. 2009. Super wicked problems and climate change: Restraining the present to liberate the future, *Cornell Law Review* 94 (5): 1152–234.

Le Masurier, Megan. 2015. What is slow journalism?, *Journalism Practice* 9 (2): 138–52.

Lewis, Simon L. and Mark A. Maslin. 2015. Defining the Anthropocene, *Nature* 519 (7542): 171–80.

Lorimer, Hayden. 2005. Cultural geography: The busyness of being 'more-than-representational', *Progress in Human Geography* 29 (1): 83–94.

Lorimer, Jamie. 2015. *Wildlife in the Anthropocene: Conservation After Nature.* University of Minnesota Press.

Malm, A. and A. Hornborg. 2014. The geology of mankind? A critique of the Anthropocene narrative, *Anthropocene Review* 1 (1): 62–9.

Marx, Karl and Friedrich Engels. 2002. *The Communist Manifesto*, ed Gareth Stedman Jones. Penguin.

May, Theresa. 2016. Speech to the Conservative Party Conference [online]. 5 October. Available at: www.ukpol.co.uk/theresa-may-2016-speech-at-conservative-party-conference/ (Accessed 6 October 2025)

Mazzucato, M. 2011. The entrepreneurial state, *Soundings* 49 (49): 131–42.

McCartney, G., C. Collins, D. Walsh and G.D. Batty. 2012. Why the Scots die younger: Synthesizing the evidence, *Public Health* 126 (6): 459–70.

McKibben, Bill. 2005. The emotional core of *The End of Nature, Organization & Environment* 18 (2): 182–5.

McKinlay, Alan. 2002. 'Dead selves': The birth of the modern career, *Organisation* 9 (4): 595–614.

McLuhan, Marshall. 1964. *Understanding Media: The Extensions of Man*. McGraw-Hill.

McLuhan, Marshall. 1974. At the moment of Sputnik the planet became a global theatre in which there are no spectators but only actors, *Journal of Communication* 24 (1): 48–58.

Meyer, Robinson. 2016. Donald Trump is the First Demagogue of the Anthropocene, *The Atlantic*. Available at: www.theatlantic.com/science/archive/2016/10/trump-the-first-demagogue-of-the-anthropocene/504134/ [Accessed 22 September 2025].

Moore, Jason W. 2017. The capitalocene, part I: On the nature and origins of our ecological crisis, *Journal of Peasant Studies* 44 (3): 594–630.

Morley, David. 2017. *Communications and Mobility: The Migrant, the Mobile Phone, and the Container Box*. Wiley-Blackwell.

Morton, Timothy. 2013. *Hyperobjects: Philosophy and Ecology after the End of the World*. University of Minnesota Press.

Moser, Susanne C. and Lisa Dilling. 2011. Communicating climate change: Closing the science–action gap, in John S. Dryzek, Richard B. Norgaard and David Schlosberg (eds), *The Oxford Handbook of Climate Change and Society*. Oxford University Press, pp 161–74.

Nowviskie, B. 2015. Digital humanities in the Anthropocene, *Digital Scholarship in the Humanities* 30 (suppl_1): i4–i15.

O'Hara, Kieron. 2020. The contradictions of digital modernity, *AI & Society* 35 (1): 197–208.

Örnebring, Henrik and Amy Schmitz Weiss. 2021. Journalism and the politics of mobility, *Journalism Studies* 22 (14): 1894–910.

Owen, Laura Hazard. 2021. Facebook's pivot to video didn't just burn publishers. It didn't even work for Facebook. NiemanLab, 15 September. Available at: www.niemanlab.org/2021/09/well-this-puts-a-nail-in-the-news-video-on-facebook-coffin/ [Accessed 22 September 2025].

Painter, James, María Carmen Erviti, Richard Fletcher, Candice Howarth, Silje Kristiansen, Bienvenido León, Alan Ouakrat et al. 2016. *Something Old, Something New: Digital Media and the Coverage of Climate Change*. Reuters Institute for the Study of Journalism.

Parikka, Jussi. 2016. Deep times of planetary trouble, *Cultural Politics* 12 (3): 279–92.

Parks, Lisa. 2005. *Cultures in Orbit: Satellites and the Televisual*. Duke University Press.

Parks, Perry. 2017a. Silent spring, loud legacy: How elite media helped establish an environmentalist icon, *Journalism & Mass Communication Quarterly* 94 (4): 1215–38.

Parks, Perry. 2017b. Non-representational news: An intervention into pseudo-events, *Journalism* 22 (1): 122–38.

Philo, Greg and Catherine Happer. 2013. *Communicating Climate Change and Energy Security: New Methods in Understanding Audiences*. Routledge.

Pincetl, Stephanie. 2017. Cities in the age of the Anthropocene: Climate change agents and the potential for mitigation, *Anthropocene* 20: 74–82.

Press Gazette. 2011. George Monbiot reveals Guardian wage and urges other journalists to do the same. *Press Gazette*, 29 September. Available at: https://pressgazette.co.uk/george-monbiot-revealsguardian-wage-and-urges-other-journalists-to-do-the-same/ [Accessed 22 September 2025].

Pringle, Tom. 2020. The Climate Proxy: Digital Cultures of Global Warming. PhD thesis, Brown University.

Rosling, Hans, Ola Rosling and Anna Rosling Rönnlund. 2018. *Factfulness: Ten Reasons We're Wrong about the World – and Why Things are Better than you Think*. Flatiron Books.

Sachsman, David B. and JoAnn Myer Valenti (eds). 2020. *Routledge Handbook of Environmental Journalism*. Routledge.

Scranton, Roy. 2015. *Learning to Die in the Anthropocene: Reflections on the End of a Civilisation*. City Lights.

Searle, Adam, Jonathon Turnbull and Jamie Lorimer. 2021. After the anthropause: Lockdown lessons for more-than-human geographies, *Geographical Journal* 187 (1): 69–77.

Sheller, Mimi and Urry, John. 2006. The new mobilities paradigm, *Environment and Planning A* 38 (2): 207–26.

Sklair, Leslie (ed). 2020. *The Anthropocene in Global Media: Neutralizing the Risk*. Routledge.

Skrimshire, Stefan. 2018. Confessing Anthropocene, *Environmental Humanities* 10 (1): 310–29.

Skrydstrup, Martin. 2013. Tricked or troubled natures? How to make sense of 'climategate': Responding to the Challenges of our Unstable Earth (RESCUE), *Environmental Science & Policy* 28: 92–9.

Sobel Fitts, Alexis. 2014. Narrating climate change. *Columbia Journalism Review*, 19 June.

Star, Susan Leigh and James R. Griesemer. 1989. Institutional ecology, 'translations' and boundary objects: Amateurs and professionals in Berkeley's Museum of Vertebrate Zoology, 1907–39, *Social Studies of Science* 19 (3): 387–420.

Steffen, Will, Wendy Broadgate, Lisa Deutsch, Owen Gaffney and Cornelia Ludwig. 2015. The trajectory of the Anthropocene: The Great Acceleration, *Anthropocene Review* 2 (1): 81–98.

Stephens, Lucas, Erle Ellis and Dorian Fuller. 2020. The deep Anthropocene. *Aeon*, 1 October. Available at: https://aeon.co/essays/revolutionary-archaeology-reveals-the-deepest-possible-anthropocene [Accessed 22 September 2025].

Sutton Trust. 2019. *Elitist Britain 2019: Summary Report*. Sutton Trust. Available at: www.suttontrust.com/wp-content/uploads/2020/01/Elitist-Britain-2019-Summary-Report.pdf [Accessed 22 September 2025].

The Economist. 2011. Welcome to the Anthropocene, *The Economist*, 28 May, sec. 399.

Thompson, E.P. 1967. Time, work-discipline, and industrial capitalism, *Past & Present* 38 (1): 56–97.

Tsing, Anna. 2000. The global situation, *Cultural Anthropology* 15 (3): 327–60.

United Nations. 2024. World population prospects 2024: Summary of results. UN Department of Economic and Social Affairs. Available at: https://population.un.org/wpp/ [Accessed 22 September 2025].

Usher, Nikki. 2021. *News for the Rich, White, and Blue: How Place and Power Distort American Journalism*. Columbia University Press.

Van Leuven, Sarah and Peter Berglez. 2016. Global journalism between dream and reality: A comparative study of The Times, Le Monde and De Standaard, *Journalism Studies* 17 (6): 667–83.

Wahl-Jorgensen, Karin. 2016. Emotion and journalism, in Tamara Witschge, C.W. Anderson, David Domingo and Alfred Hermida (eds), The SAGE Handbook of Digital Journalism. SAGE, pp 128–44.

Wainwright, Joel and Geoff Mann. 2013. Climate leviathan, *Antipode* 45 (1): 1–22.

Wall, Melissa. 2015. Citizen journalism: A retrospective on what we know, an agenda for what we don't, *Digital Journalism* 3 (6): 797–813.

Wall, Melissa. 2017. Mapping citizen and participatory journalism: In newsrooms, classrooms and beyond, *Journalism Practice* 11 (2–3): 134–41.

Walsh, David, Gerry McCartney, Chik Collins, Martin Taulbut and George D. Batty. 2017. History, politics and vulnerability: Explaining excess mortality in Scotland and Glasgow, *Public Health* 151: 1–12.

Weston, Phoebe. 2020. Mass die-off of birds in south-western US 'caused by starvation', *The Guardian*, 26 December. Available at: www.theguardian.com/environment/2020/dec/26/mass-die-off-of-birds-in-south-western-us-caused-by-starvation-aoe [Accessed 22 September 2025].

Widholm, Andreas. 2019. Transnational news consumption and digital content mobility: Insights from Sweden, *Journalism Studies* 20 (10): 1472–90.

Wimmer, Andreas and Glick Schiller, Nina. 2002. Methodological nationalism and beyond: nation–state building, migration and the social sciences, *Global Networks* 2 (4): 301–34.

Yusoff, Kathryn. 2018. *A Billion Black Anthropocenes or None*. University of Minnesota Press.

Zalasiewicz, Jan, Mark Williams, Will Steffen and Paul Crutzen. 2010. The new world of the Anthropocene, *Environmental Science & Technology* 44 (7): 2228–31.

Zelizer, Barbie. 2021. Why journalism's default neglect of temporality is a problem, *Media, Culture & Society* 43 (7): 1213–29.

Zelizer, Barbie. 2022. What journalism tells us about memory, mind and media, *Memory, Mind & Media* 1: e6.

Zevin, Alexander. 2021. *Liberalism at Large: The World According to The Economist*. Verso Books.

Index

References to figures are in *italic* type; those in **bold** type refer to tables. References to endnotes show the page number, note number and chapter number (132n2(ch5)).

1980s 5, 9, 15–16
1990s 14, 15–16, 36, 113
2 degrees Celsius target 36
2000s 16–17, 42
3D visualisation 43

A

abundance 8–9
accelerationist approaches 112
accountability of news media 93
activism, climate 66–67, 92, 93
actor–network theory (ANT) 13, 116
Adam, Barbara 87, 129
adaptability 98
  and climate change 74, 75, 90
  and journalism 94, 112, 127
Afghanistan 46
agency in journalism 7, 9, 120, 129
agriculture **23**
air travel 48, 63, 77, 101
alarmism 59
Alexievich, Svetlana, *Chernobyl Prayer* 34
Amin, Ash and Thrift, Nigel 76
Anderson, Ben 89
Anderson, Benedict R.O.G. 7
Anderson, Christopher 59
animals, electric 47
Anthropocene 11–12, 18–25, 31–40, 51–65, 124–127
  and capitalism 20–21, 29, 32
  and colonialism 21, 31, 82
  and communications technology 21, **22**, 24, 49
  and cosmopolitanism 29, 100–101, 106, 108, 109–111
  and data collection and analysis 52–53, 56–57
    and modernity 51, 52, 64–65
    and wealth 61, 62, 63
  and environmental issues 18–19, 47, 52, 67, 109–110
  and carbon emissions 20, **23**, 25, 65, 85
  and climate change 20, 21, **23**, 35–36, 38, 61, 123
  and everyday life 36, 39–40, 124–127
  and globalisation 19, **22**, 31
  and Great Acceleration 4, 20, 21, **23**
  and infrastructure 37, 38–39, 52
  and meaning 36, 37–39, 40, 50
  and modernity 4, 5–6, 25, 37–39, 51–65, 87, 131
    and data collection and analysis 51, 52, 64–65
  and news media 39–40, 53
  and power 52, 53
  and representation 124–125, 127
  and storytelling 49, 51, 125
  and time 38, 81, 85, 86–88, 128–129
  and trends 21, **22**, **23**
  and urbanisation **22**, **23**, 67, 68, 77–78
  see also cities; liquidity/liquid journalism; resilience
Anthropocene archetypes 33–36
Anthropocene events 38, 39
Anthropocene journalism 12, 28–29, 33, 49–50, 125–126, 128
Anthropocene ocean 47
anti-rationalism 132n2(ch5)
Atlantic journal 32
attraction, journalistic 33–34
audiences
  and cosmopolitanism 106, 109
  and news media 91, 92, 97, 106
augmented reality (AR) 43
Australian news sites 95
authoritarianism, democratic 101

B

Baldwin, Andrew 49
banality 102, 110, 114
  see also everyday life

Bangladesh 68, 117
barriers, breaking 43, 81
Bauman, Zygmunt 112, 114–115
BBC 3, 10, 38
Beast, The 66
Beck, Ulrich 15, 25, 100, 101, 102, 122
Beijing 68
Bell, Karen 39
Berlant, Lauren 60–61, 86, 130
Berman, Marshall 5, 69
berry picking, Chernobyl 35
Bezos, Jeff 29
Bhambra, Gurminder K. 101
Biden, Joe 66, 72
biodiversity 73
biosphere 21, **23**, 52
bird populations 25, 132n1(ch2)
blogging 105
Boorstin, Daniel 125
Borges, Jorge Luis 54
boundary objects 56
branding 92, 103–104
Britain/United Kingdom 27, 48, 64, 72
 and cosmopolitanism 101, 103–104, 109
 and environmental issues 27, 117
 and liquidity/liquid journalism 115, 117
British Columbia 123
Brown, Kate 35
buildings, historic, Glasgow 74
Bulletin of the Atomic Scientists 83–84
Burawoy, Michael 115
Buzzfeed 91–92

C

cable TV 9–10
Calhoun, Craig 99
Callison, Candis and Young, Mary Lynn 124
cancer 57
capitalism 9, 69–70
 and Anthropocene 20–21, 29, 32
 and change 5, 118, 132n1(ch10)
 and colonialism 21, 82
 and cosmopolitanism 82, 101, 103–104
 and data collection and analysis 59, 60, 62
 and environmental issues 5, 16, 17–18, 20–21
   and climate change 5–6, 29, 59, 60, 113
 and future time 96–97, 123, 124, 128, 129–130
 and globalisation 10, 11, 62
 and modernity 5, 6, 39, 40, 91, 118, 124, 132n1(ch10)
 and newspapers 6, 8
 and resilience 91, 93–94, 95
 and United States 18, 21
capitalism, digital 29, 91, 95
capitalism, extractive 5, 32, 82
capitalism, global 62, 128

capitalism, industrial 6, 17–18, 69–70
capitalism, liberal 16
capitalism, media 5, 11, 91, 93–94, 129–130
car dependency **22**, 70, 71, 72, 82
carbon emissions
 and Anthropocene 20, **23**, 25, 65, 85
 and cities 68, 75, 76
   Glasgow 71, 73, 74
 and concrete 20, 72, 75, 77–78
 and future time 85, 113
 and Great Acceleration 20, **23**
 and IPCC 84, *85*
 and low-carbon/carbon zero technologies 73, 74, 75
 and modernity 65, 113
carer role 105
cars, electric 88
Carson, Rachel, *Silent Spring* 17–18, 25, 47
change 32
 and capitalism 5, 118, 132n1(ch10)
 and liquidity 112, 113
 and modernity 5–6, 112, 118, 132n1(ch10)
 see also environmental issues
change, radical 32
change, social 4, 69–70, 71–72, 75, 126, 127
change, technological 4, 9, 12, 50, 96, 119–120
 and future time 126–127, 128
changes, editorial and commercial 110
Chernobyl 33–35
China 21, 33, 115
Christophers, Brett 97
cities 66–78
 and climate change *55*, 70–71, 77–78
   and carbon emissions 68, 71, 73, 74, 75, 76
 and everyday life 73, 75–76, 77, 78
 and infrastructure 28, 37, 71–72, 73, 75, 77
 and networks 72–73, 75–77
 and pollution 68, 70, 78
 and post-industrialism 68–69, 72, 74–75
 and present time 68, 69
 and public goods 60, 74
 and suburban areas 72, 77
 and sustainability 67, 70, 73, 74
 see also Glasgow
cities journalism 67–68, 69, 78
Cities project, *Guardian* newspaper 68
citizen journalism 105
Citylab 68
Clarke, Arthur C. 43
Climate Central platform 56
climate change 14–15
 and adaptability 74, 75, 90
 and Anthropocene 20, 21, **23**, 35–36, 38, 61, 123
 and capitalism 5–6, 29, 59, 60, 113
 and cities *55*, 68, 70–71, 77–78
   and carbon emissions 68, 71, 73, 74, 75, 76

INDEX

and communications technology 21, 45, 46, 48, 49
and cosmopolitanism 103–104, 106–108
and data collection and analysis 54–56, 57, 58–64, 116
   and fossil fuels 58, 61
   and IPCC 54, 61
   and modernity 60, 61–62, 65
   and politics 58, 59, 60
   and rigour, scientific/academic 58, 59
   and storytelling 61–64, 65
and flooding 37, 45, 70–71, 73, 108, 123
and global economy 39, 108
and global warming 36, 70, 71, 77–78
and innovation 29, 60, 113
and liquidity/liquid journalism 113, 114–115, 117–118
and migration 107, 117–118
and modernity 5–6, 113, 123–124, 131
   and data collection and analysis 60, 61–62, 65
and news media 84, 91
and reality 125, 126–127
and resilience 90, 91, 92–93
and Scotland 37–38, 70–71
and sea levels 56, 70–71, 113
and time 83, 84–85, 86, 113, 122–124, 127–130
and transnationalism 106–107, 125
and weather 37–38, 45, 70–71
see also carbon emissions; environmental issues
climate change denial 22, 58–59
climate journalism 92, 93
climate refuges 70, 132n1(ch6)
Climatic Research Unit, UEA 58, 59
CNN 9, 10
coastal pollution **23**
colonialism 20, 21, 31, 82, 117
   European colonialism 25, 80, 117–118
Columbia University 115
communications technology 2–3, 9, 41–50
and Anthropocene 21, **22**, *24*, 49
and environmental issues 21, 44–46, 47, 48
and immediacy 46, 48
and meaning 46, 50
and modernity 2–3, 7, 47–50
and news media 9, 43, 45, 47, 91, 104
and reality 43, 47
and smartphones 42, 44, 45
and social media 45, 46, 47
and sound 43–45, 46–47
and storytelling 43–44, 49
and technological change 4, 9, 12, 50, 96, 119–120
   and future time 126–127, 128
   and tools of journalism 42, 43, 44
community 97
community news organisation 73
companies, media 10, 11, 89, 92, 95, 126

companies, multinational 10, 103
companies, private 63
companies, technology 11, 95
computer-assisted reporting (CAR) 53–54
computer-generated imagery (CGI) 43
concrete 20, 72, 75, 77–78
connectivity/interconnectiveness 65, 76, 110
   see also networks
consequence, chains of 64
consequences, real world 54, 88
conspiracy theories 7, 58–59
corporate planning 92
Correspondent digital news site 92
cosmopolitan journalism 102, 103, 108
cosmopolitanism 76, 99–111
and Anthropocene 29, 100–101, 106, 108, 109–111
and audiences 106, 109
and Britain/United Kingdom 101, 103–104, 109
and capitalism 82, 101, 103–104
and environmental issues 103–104, 106–108, 109–110
and everyday life 102, 110
and foreign correspondents 108, 110
and globalisation 99–106, 109
and inequality 100, 104
and journalistic mobility 100, 101
and migration 107, 108–109
and modernity 99, 108, 110
and nationalism 101, 103, 108–109, 110–111
and networks 72–73, 75–77
New York Times 103, 105–106
sociology/theory of 100–101, 102
and transnationalism 99, 106–107, 108–109
and utopianism 100, 102, 110
cosmopolitanism, American 106
cosmopolitanism, Anthropocene 109–111
cosmopolitanism, archetypal 99, 103, 110
cosmopolitanism, banal/everyday 102, 110
cosmopolitanism, economic 99
cosmopolitanism, elite 101–102, 103, 110
cosmopolitanism, European 99–100, 101
Cottle, Simon 109
COVID-19 39, 42, 63, 72, 119
creativity of journalists 118, 119, 120
Cronon, William 26
Crossrail project, London 77
crowdfunding of journalism 94, 96, 120
Crutzen, Paul 18, 51
curators, journalists as 114

D

dam building 22
data, charismatic 64–65
data collection and analysis 51–65, 66–78, 132n1(ch5), 132n2(ch5)
and Anthropocene 52–53, 56–57

145

and modernity 51, 52, 64–65
and wealth 61, 62, 63
and capitalism 59, 60, 62
and commercial information 63, 132n3(ch5)
and data journalism 53, 64
and environmental issues 52, 54–57, 58–64, 116
and fossil fuel industry 58, 61, 62, 63
and IPCC 54, 61
and modernity 60, 61–62, 65
and politics 58, 59, 60
and scientific/academic rigour 58, 59
and storytelling 61–64, 65
and global financial system 52, 62, 64
and governance 60, 63
and hacking 58, 62
and hyperobjects 56–57, 64
and information gathering/knowledge production 51, 52, 53, 60
and modernity 51, 52, 60, 61–65
and news media 53, 59, 61
and power 52, 53, 59, 63
and rationality 53, 59
and social media 56, 62
and statistics 57, 59–60, 61, 63
and storytelling 51, 53, 61–64, 65
and trends 54–55, 61
and visualisation 55–56, 61
see also climate change
data journalism 51, 53–54, 56, 62, 64
decision-making 53
deforestation **23**
Delanty, Gerald 101
democracies, liberal 26, 94, 109
democracy 3, 73, 86, 105
Demos, T.J. 19
deregulation 10, 112
Desrosières, Alain 132n1(ch5)
Deuze, Mark 7, 49, 114, 120
development, neoliberal 68, 69
development, private 70, 73–74
development, urban 88
Dickens, Charles 132n1(ch3)
digital native journalism 90–92
disasters, natural 37–38, 45, 90
discussion, academic 31
disembedding/re-embedding 119–120
disintegration 112, 116–118, 121
disruption 90–93
domestic journalism 108
Donovan, Kevin P. 86
drought 108
Dryzek, John S. 97

E

earth system trends 21, **23**
Ecologist journal 17

ecomodernist journalism 113
economic cosmopolitanism 99
economic decline, Glasgow 68–69
economic security 75, 120
economic system, global 39
economies
and climate change 39, 75, 108
Glasgow 72, 73, 74, 75
global economies 9, 39, 97, 108
Economist journal 32, 53
ecosystem services 76, 131
editorialisation 58
education of journalists 27
see also training, professional
efficiency 42, 60
Ellis, Erle 19
employment opportunities 28, 74–75
see also labour of journalists/working life of journalists
end of journalism 13–30
energy, renewable 37, 75, 76, 88
energy use **22**, 74
Energydesk 93
Eno, Brian 31
entertainment industry 8
entrepreneurship, journalism 96, 118–120
environmental issues 14–30
and Britain/United Kingdom 27, 117
and capitalism 16, 17–18, 20–21
and climate change 5–6, 29, 59, 60, 113
and colonialism 20, 21, 117
and communications technology 21, 44–46, 47, 48
and cosmopolitanism 103–104, 106–108, 109–110
and data collection and analysis 52, 54–57, 58–64, 116
and modernity 60, 61–62, 65
and politics 58, 59, 60
and scientific/academic rigour 58, 59
and storytelling 61–64, 65
and fossil fuel industry 58, 61, 62, 63, 66, 103–104
and nature 14–25, 27–28, 29–30
and news media 16, 84, 91
and newspapers 27, 28, 38, 39
and Scotland 27–28, 37–38, 70–71
and United States 14, 17, 21, 25, 26, 79–80
see also Anthropocene; carbon emissions; climate change; Intergovernmental Panel on Climate Change (IPCC)
environmental journalism 14, 15–18, 25–26, 27–28, 29–30
environmentalism, elite 26, 27
environmentalism, working-class 39–40
escatology, journalism as 122–131
see also time, future
ethics of journalism 8, 16, 53, 110

eurocentrism 101
European Commission 100
European Union 100
everyday life 65, 86
   and Anthropocene 36, 39–40, 124–127
   and cities 73, 75–76, 77, 78
   and cosmopolitanism 102, 110
   and liquidity 114, 116, 121
exclusion zones 34
Exxon 93, 132n3(ch5)
eyewitness accounts 105

F

'factfulness' 59
failure, systematic 64
farming, urban, Glasgow 73
Farrier, David 80
fertiliser use **22**
fieldwork, journalistic 35, 48, 50
file sharing 62
financial analysis 53
financial crisis (2008) 64, 96
financial journalism 53
financial system, global 52, 62, 109
Financial Times 92
financialisation 10, 71, 95, 96, 97
fishing industry **23**
flexibility, journalistic 48, 94, 112–113, 120
flooding 37, 45, 70–71, 73, 108, 123
flow and disintegration 116–118, 121
food supplies 72–73, 76, 109
foreign correspondents 8, 108, 110
foreign direct investment **22**
forest fires/wildfires 45, 46, 123
fossil fuel industry 58, 61, 62, 63, 66, 103–104
freedom, human 96
freelancers 17, 53, 104, 113, 119–120
Fukushima nuclear power plant 35
Fukuyama, Francis 16
future planning, Glasgow 71–75
futures market 93–97
futurology 29

G

Gabrys, Jennifer 45
Garnett, Philip 64
Gasher, Mike 83, 128–129
GDP **22**
gentrification, artistic, Glasgow 69
geopolitics 32, 66
German Rhineland 6
Germany 100, 108
Ghosh, Amitav 130
glaciers 79–80
Glasgow 66–78
   and climate change 55, 70–71, 73, 74
   and development, neoliberal 68, 69
   and development, private 70, 73–74
   economy of 68–69, 72, 73, 74, 75
   and everyday life 73, 75, 78
   and industrialisation 68–69, 74–75
   and innovation 72, 75
   and investment 74, 75
   and networks 76, 77
   and post-industrialism 68–69, 72, 74–75
   and public transport 70, 71, 77
   and resilience 73, 74
   and UN COP 66–67, 68–69, 72–73, 103, 123, 125
   and urban regeneration 70, 71, 73
   and utopianism 70, 75
Glasgow effect 68, 78
Glasgow Herald newspaper 69
global journalism 10, 11, 48, 112–121
   and cosmopolitanism 99, 102–106
   and *Our World* 2, 56
Global North 27, 32, 101, 107, 109
global warming 36, 70, 71, 77–78
globalisation
   and Anthropocene 19, **22**, 31
   and capitalism 10, 11, 62
   and cosmopolitanism 99–106
globality 2, 19, 95, 99, 101, 102
good intentions 87–88
Google 43
governance 60, 63, 87–88, 95–96
governance, global 16, 103, 126
Graham, Phil 29
Great Acceleration 4, 18, 20, 21, **23**, 48, 59
Great Lakes, America 57
great man theory 64
Greenpeace 93
Gregory, Karen 52
Guardian newspaper 41, 68, 103

H

Habermas, Jürgen 100
habitat loss **23**, 39
hacking 58, 62
Hamilton, Clive 59
Hammarskjöld, Dag 95
Hannerz, Ulf 103
Happer, Catherine 86, 96
Haraway, Donna 20, 87
Hartley, John 3, 6, 7, 40, 94
Hauptmann, Gerhart 132n1(ch3)
heatwaves 123, 126
Hersey, John, *Hiroshima* 35
Hesmondhalgh, David and Baker, Sarah 118
Hiroshima 35
history, rewriting of 81–82, 117
hobbyism 105, 106
housing, Glasgow 73–74
human development 4

human influence on nature 13, 14, 15–16, 18–25
human interest stories 116
Human Interference Task Force 80
human-made materials 20
human rights 94–95
humanitarian assistance 107
Hunter, Andrea 120
hybrid formats 43–45
hyperobjects 56–57, 64

I

ideology 58
imaginary, sociotechnical 127–130
immediacy 10, 46, 48, 123
inclusion 102
independence, editorial 93
India 68
industrialisation **22**, 37, 68–69, 74–75, 117
industries, green 75
inequality 100, 104, 124
information, cartelisation of 95
information, commercial 63, 132n3(ch5)
information, freedom of 94
information society 114
information transmission 49
infrastructure 46
  and Anthropocene 37, 38–39, 52
  and cities 28, 37, 71–72, 73, 75, 77
Ingold, Tim 46
innovation 4, 32
  and climate change 29, 60, 113
  Glasgow 72, 75
  and reality 126, 128, 129
  and resilience 89, 92, 94, 96
institutionalism, new 97–98
institutions, democratic 16, 131
institutions, media 96, 97–98
institutions, public 90, 98
institutions, transnational 93, 95
integrity of science 58
intellectualism 103
Intergovernmental Panel on Climate Change (IPCC) 15, 36, 106–107, 123
  and carbon emissions 84, *85*
  and data collection and analysis 54, 61
International Commission on Stratigraphy 51
International Consortium of Investigative Journalists 62
international development 59
International Federation of Red Cross and Red Crescent Societies (IFRC) 107
International Geosphere–Biosphere Programme (IGBP) 21
International Union of Geological Sciences (IUGS) 18–19
internationalisation 109
internet 10, 41

invention 25–28, 29
investigative journalism 92, 93
investment 74, 75, 93, 97

J

Jacobs, Jane 71
Jamie, Kathleen 26, 27
Japan 35, 68
Jasanoff, Sheila 127–128
journalism *see* Anthropocene; capitalism; cities; cosmopolitanism; data collection and analysis; individual types of journalism; modernity; resilience; time
journalistic methods 108–109
judgement, editorial and curatorial 49, 111

K

Kanngieser, Anja 46
Kant, Immanuel 100, 102
'Keeling curve' 65
'Keep it in the Ground' campaign 103
Kitch, Carolyn 81
knowledge production 18, 60, 65
Kolbert, Elizabeth 35–36

L

labour, flexible 11, 17, 48, 120
labour, free 105
labour of journalists/working life of journalists 11, 17, 48, 104–106
  freelancers 17, 53, 104, 113, 119–120
  see also liquidity/liquid journalism
Langston, Nancy 57
language 80
laptops 42, 44, 119
Latour, Bruno 13–14, 116, 122
Le Monde 92
liberalism 16, 26, 31
Library of Babel (Borges) 54
life expectancy 68
lifestyle 27, 60, 126
liquidity/liquid journalism 112–121
  and Britain/United Kingdom 115, 117
  and change 112, 113
  and climate change 113, 114–115, 117–118
  and disintegration 112, 116–118, 121
  and everyday life 114, 116, 121
  and flow and disintegration 116–118, 121
  and future time 112, 113, 121
  and modernity 113, 114–116, 119–120, 121
  and professional training/skills 114–116, 121
living standards 83
London 67, 77, 123
Lorimer, Hayden 125
Lorimer, Jamie 47
Los Angeles 68

INDEX

M

M8 motorway, Glasgow 71–72
Macfarlane, Robert 27
Malm, Andreas 19
managerialist approaches 112
Manhattan project 35
manipulation of data 58
mapmaking 83, 128–129
Mariana Trench 123
marketing 31
Marx, Karl 5, 6
Mauna Loa Observatory, Hawaii 65
May, Theresa 101
Mazzucato, Mariana 97–98
McKibben, Bill 13–14, 26, 122
McKinlay, Alan 118
McLuhan, Marshall 1, 2, 44, 56
meaning 36, 37–39, 40, 46, 50, 122
media companies 11, 89, 92, 95, 126
media departments, university 42
media, elite 32, 39, 43, 106
media, global 9–11, 56, 106, 116
'media life' 7–8
media networks 11, 45–46, 56, 119–120
media, nonprofit 92
media platforms 41, 90–93, 119
media, wildfire 45–47
memory 81–82
methane **23**
migration 76, 107, 108–109, 117–118
Millennium Development Goals 84
mining 33, 88
mobility, journalistic 47–49, 100, 101
modernisation, Glasgow 68–69
modernity 3–9, 11–12, 51–65, 130–131
  and Anthropocene 4, 5–6, 25, 32, 33, 37–39, 87, 131
    and data collection and analysis 51, 52, 64–65
  and capitalism 5, 6, 39, 40, 91, 118, 124, 132n1(ch10)
  and change 5–6, 112, 118, 132n1(ch10)
  and climate change 5–6, 60, 61–62, 65, 113, 123–124, 131
  and communications technology 2–3, 7, 47–50
  and cosmopolitanism 99, 108, 110
  and data collection and analysis 51, 52, 60, 61–65
  and expectation 60–61, 113
  and liquidity/liquid journalism 113, 114–116, 119–120, 121
  and nature 18, 20, 21, 25, 26
  and sense-making 3, 6, 7, 40, 88, 124, 131
  and sociological thought 13, 114–116
  and storytelling 49, 64
  and time 85, 87, 113, 123–124
modernity, capitalist 39, 40, 91, 108

modernity, digital 51–65
  see also data collection and analysis
modernity, first 15
modernity, global 1–3, 110
modernity, hallmarks of 112
modernity, industrial 20, 117
modernity, liquid 112, 113, 119–120, 121
modernity, second 15
modernity, zombie 124
Moore, Jason 20
Morley, David 76
Morton, Timothy 56–57
motorways 71–72
multimedia journalism 44
music 31, 96–97

N

narcissism 31
National Centre for Atmospheric Science 55
nationalism 101, 103, 108–109, 110–111
  see also transnationalism
nations 7, 106–107
naturalism 46
nature 14–30, 109–110
  and Anthropocene 18–25
  and environmental issues 14–25, 27–28, 29–30
  and human influence 13, 14, 15–16, 18–25
  and modernity 18, 20, 21, 25, 26
  and rewilding 34, 82
  and risk management 15, 16
  and United States 14, 25, 26
nature documentaries 47
nature journalism 27, 29
neoliberalism 68, 69, 101, 119
networks 72–73, 75–77, 109
  see also communications technology; media networks
New York Times 43, 67, 92, 103, 105–106
New Yorker magazine 14, 35
News Corp 10
news, digital 47, 91, 104
news events 38–39, 87
news, fake 7–8, 59, 60
news, good 59, 61
news media 3, 17
  and Anthropocene 39–40, 53
  and audiences 91, 92, 97, 106
  and communications technology 9, 43, 45, 47, 91, 104
  and data collection and analysis 53, 59, 61
  and environmental issues 16, 84, 91
  and resilience 90–93, 94, 96
  and time 84, 86, 87, 96
news media, elite 106
news organisations *see* news media
news, 'pop-up' 94

newspapers 41, 69
  and capitalism 6, 8
  and environmental issues 27, 28, 38, 39
NGOs 92, 93
Nikkei 10
nitrous oxide **23**
nondigital journalism 44
Nordhaus, William 36
Nordic countries 115
North America 7, 27, 48, 109
novelty/newness 4, 31, 91, 94
Nowviskie, Bethany 63
nuclear conflict 2, 83–84
nuclear technology 33–35, 79–81

O

Obama, Barack 125
objectivity 53, 58
ocean acidification **23**
offshoring, financial 62
online journalism 50, 105
  see also communications technology
open commons 96
open source data 62–63
opera 31
optimism 60–61, 86, 92, 130
otherness 102
Our World TV programme 1–2, 3, 46, 48, 56, 110
ownership 10, 95, 97
ozone, stratospheric **23**

P

Pakistan 45, 68
Panama Papers scandal 62
pandemics 39
paper production **22**
Parikka, Jussi 83
Paris Agreement 36, 123
parks, Glasgow 71
Parks, Lisa 2
Parks, Perry 125
participation 95, 104–106
participatory condition 2, 105
paternalism 32, 102–104, 124, 127
Patsy role 105
Philo, Greg and Happer, Catherine 117
pilot projects 49
Pincetl, Stephanie 67
place 46–47
Plutonium-239 80
political influence 4–5
politics 16, 58, 59, 60, 125, 126
  geopolitics 32, 66
pollution **23**, 57, 68, 70, 78, 81, 123
population control 107
population growth 2, **22**, 67, 68
post-humanism 33, 34, 116

post-industrialism 37, 68–69, 72, 74–75
postmodernism 132n2(ch5)
power 52, 53, 59, 63
power, cultural 26
power relationships 105–106
precarity 86, 120
Pringle, Thomas 46
Pripyat, Ukraine 34
private sector 63, 70, 73–74, 95, 98
problems, wicked 87–88
professional journalism 46, 53, 105, 109, 115
  semi-professional 105, 106
profitability 11, 89, 91, 92, 95, 96–97
progress, human 2–3, 32
Project Iceworm 79–80
proximity 108
pseudo events 125, 127
public goods 60, 74, 93–97, 131
public service journalism 8, 10, 49, 92

Q

qualitative journalism 60
quality 91, 92, 93–94

R

radio 42
railways 37–38, 77, 132n1(ch3)
rainfall 37, 62, 70
rationality 53, 59
Reading FC 56
realism 60, 116, 125–126, 127, 129
realism, ecological 116
reality 7–8, 43, 47, 102, 124–127, 128, 129–130
recycling 75
reform 91, 94, 97, 115
refugees 70, 107, 117
regeneration, urban 70, 71, 73
regulation 91, 95, 96, 97
representation 53, 95, 117, 124–125, 127
resilience 89–98
  and capitalism 91, 93–94, 95
  and climate change 90, 91, 92–93
  and Glasgow 73, 74
  and information 94, 95
  and innovation 89, 92, 94, 96
  and news media 90–93, 94, 96
  and reform 91, 94, 97
  and regulation 91, 95, 96, 97
  and survival 89, 90, 95
  and sustainability 89, 90–93
Resurgence journal 17
retrofitting of housing stock 74
Revkin, Andrew 28
rewilding 34, 82
rigour, scientific/academic 59
Rio Earth Summit, UN 15
risk management 15, 16

Robinson, Kim Stanley, *Ministry for the Future* 132n1(ch6)
rural areas/communities 27–28, 76
Rusbridger, Alan 103
Russia 34, 115

S

satellite technology 2, 9
Scotland
   and environmental issues 27–28, 37–38, 70–71
   and industrialisation and post-industrialism 37, 68–69, 72, 74–75
   and infrastructure 28, 37, 71–72, 73
   and renewable energy 37, 75, 76
   see also Glasgow
Scottish government 66
Scottish Highlands 76
Scranton, Roy 25, 130
sea levels 56, 70–71, 113
sense-making, journalism as 9, 95, 128
   and modernity 3, 6, 7, 40, 88, 124, 131
sensor journalism 44–46
sensory filter, journalism as 50
service provision 76
Sheller, Mimi and Urry, John 47–48, 49, 101
shipping 63
Show Your Stripes campaign 55–56
shrimp production 23
skills of journalism 104, 109, 114–116, 121
Sklair, Leslie 32
Skrimshire, Stefan 80
slow journalism 14
smart cities 60
smartphones 42, 44, 45
social class 27, 39–40, 72, 86
social media 45, 46, 47, 56, 62, 105
social progress 61, 113
social theory 87
societal influence 5
Society of Environmental Journalists 17
socio-economic trends 21, **22**
sociological thought 13, 114–116
sound 43–45, 46–47
South Korea 77
speculative journalism 81
sponsorship 92, 120
Sputnik 1, 2
statistics 57, 59–60, 61, 63
status, economic 86
steel production 75
Steffen, Will 20, 21
stock markets 53–54
Stoermer, Eugene 18
stories, human interest 116
stories, ongoing/unresolved 33–36
storytelling 14, 69–77, 122
   and Anthropocene 49, 51, 125

and communications technology 43–44, 49
and data collection and analysis 51, 53, 61–64, 65
and modernity 49, 64
storytelling, digital/immersive 43–44
streaming services 96–97
subscriptions, digital 41, 96
suburban areas 72, 77
Süddeutsche Zeitung newspaper 62
surveillance, mass 52
survival 86, 87, 89, 90, 95
sustainability 88
   and cities 67, 70, 73, 74
   and resilience 89, 90–93
Sweden 115
swimming, wild 27

T

tabloid journalism 47
Taiheiyō Belt, Japan 68
Tarkovsky, Andrei, *Stalker* 34
technosphere 52
television watching 3
Thompson, E.P. 85–87
Thunberg, Greta, *Climate Book* 56
time 79–88, 89–98
   and Anthropocene 38, 81, 85, 86–88, 128–129
   and climate change 83, 84–85, 86, 113, 122–124, 127–130
   and modernity 85, 87, 113, 123–124
   and news media 84, 86, 87, 96
   and nuclear technology 79–81, 83–84
   see also liquidity/liquid journalism; time, future; time, present
time, crisis 82–85
time, deep 80, 83, 85
time, future 60–61, 122–131
   and Anthropocene 87–88, 128–129
   and capitalism 96–97, 123, 124, 128, 129–130
   and climate change 84–85, 113, 122–123, 128
   and liquidity/liquid journalism 112, 113, 121
   and media institutions 96, 97–98
   and present time 122, 124, 125, 128, 129
   and realism/reality 124–127, 128, 129–130
   and reform 91, 94, 97
   and representation 124–125, 127
   and technological change 126–127, 128
time, geological 83
time, present 89–98
   and Anthropocene 38, 85, 87
   and cities 68, 69
   and future time 122, 124, 125, 128, 129
   and liquidity/liquid journalism 112, 113
Times newspaper 38

tools of journalism 42, 43, 44
totalitarianism 52
tourism **22**, 34
training, professional 109, 114–115, 121
transience 5–6, 94
translocalism 106, 120
transnational media corporations (TMCs) 10
transnationalism 7, 10–11, 48, 119–120
  and climate change 106–107, 125
  and cosmopolitanism 99, 106–107, 108–109
  transnational institutions 93, 95
  see also globalisation
transport, public 70, 71, 77
transportation **22**
trends 21, **22**, **23**, 54–55, 61, 89
Trump, Donald 32
Tsing, Anna 109–110, 116

U

Ukraine war 34, 46, 72
UN COP (Conference on the Parties) climate summits 15, 90
  Glasgow 66–67, 68–69, 72–73, 103, 123, 125
UN Universal Declaration of Human Rights 94
Unearthed platform 93
United Kingdom *see* Britain/United Kingdom
United Nations 16
United States
  and capitalism 18, 21
  and environmental issues 17, 21, 79–80
  and nature 14, 25, 26
  and professional training 109, 115
  and war 46, 79–80

University of East Anglia (UEA), Climate Research Unit 58, 59
uranium 80
urban journalism 67–68
urbanisation **22**, **23**, 67, 68, 77–78, 82
Usher, Nikki 27
utopianism 8, 70, 75, 100, 102, 110

V

Van Leuven, Sarah 102
visualisation, use of 55–56, 61
Vivendi 10
volunteerism 105, 106
Vonnegut, Kurt, *Slaughterhouse-Five* 79

W

Wahl-Jorgensen, Karin 7
Wainright, Joel and Mann, Geoff 126
Ward, Christopher 31
warfare 46, 79–80, 83–84
Washington Post 29
water use **22**
wealth 61, 62, 63, 71, 72, 77
weather 37–38, 45, 46, 70–71
WikiLeaks 62
wind turbines 88
working life of journalists *see* labour of journalists/working life of journalists

Y

Yucca Mountain Nuclear Waste Repository 80
Yusoff, Kathryn 19

Z

Zackenberg Research Station 123
Zelizer, Barbie 86
zero balance lives 86
Zevin, Alexander 53

www.ingramcontent.com/pod-product-compliance
Lightning Source LLC
Chambersburg PA
CBHW071712020426
42333CB00017B/2237